# 'I Follow Aristotle': How William Harvey Discovered the Circulation of the Blood

This book presents a new interpretation of how and why the discovery of the circulation of the blood in animals was made. It has long been known that the English physician William Harvey (1578–1657) was a follower of Aristotle, but his most strikingly 'modern' and original discovery – of the circulation of the blood – resulted from Harvey following Aristotle's ancient programme of investigation into animals. This is a new reading of the most important discovery ever made in anatomy by one man and produces not only a radical re-reading of Harvey as anatomist, but also of Aristotle and his investigations of animals.

**Andrew Cunningham** was for many years Wellcome Trust Lecturer and then Senior Research Fellow in the History of Medicine in the Department of History and Philosophy of Science at Cambridge University.

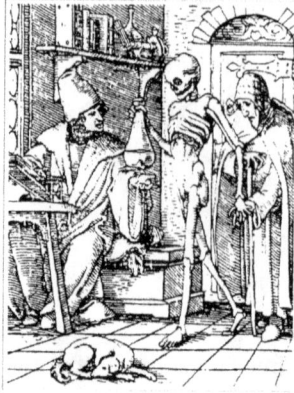

**The History of Medicine in Context**

Series Editors: Andrew Cunningham (Department of History and Philosophy of Science, University of Cambridge) and Ole Peter Grell (Department of History, Open University)

*Titles in the series include*

**Civic Medicine**
Physician, Polity, and Pen in Early Modern Europe
*Edited by J. Andrew Mendelsohn, Annemarie Kinzelbach and Ruth Schilling*

**Authority, Gender, and Midwifery in Early Modern Italy**
Contested Deliveries
*Jennifer F. Kosmin*

**Forty Days**
Quarantine and the Traveller, c. 1700–c. 1900
*John Booker*

**The World of Worm: Physician, Professor, Antiquarian, and Collector, 1588–1654**
*Ole Peter Grell*

**'I Follow Aristotle': How William Harvey Discovered the Circulation of the Blood**
*Andrew Cunningham*

For more information about this series, please visit: https://www.routledge.com/The-History-of-Medicine-in-Context/book-series/HMC

# 'I Follow Aristotle': How William Harvey Discovered the Circulation of the Blood

Andrew Cunningham

Routledge
Taylor & Francis Group

LONDON AND NEW YORK

Cover image: Portrait of William Harvey; engraving by William Faithorne, as used as frontispiece for Harvey's *Anatomical exercitations, concerning the generation of living creatures*, London, 1653. Image from Wellcome Collections.

First published 2022
by Routledge
4 Park Square, Milton Park, Abingdon, Oxon OX14 4RN

and by Routledge
605 Third Avenue, New York, NY 10158

*Routledge is an imprint of the Taylor & Francis Group, an informa business*

*British Library Cataloguing-in-Publication Data*
A catalogue record for this book is available from the British Library

*Library of Congress Cataloging-in-Publication Data*
A catalog record has been requested for this book

ISBN: 978-1-032-16223-2 (hbk)
ISBN: 978-1-032-16224-9 (pbk)
ISBN: 978-1-003-24761-6 (ebk)

DOI: 10.4324/9781003247616

Typeset in Times
by SPi Technologies India Pvt Ltd (Straive)

## Dedication: to my 'most loving collegs'

As William Harvey did, I dedicate this book to my 'most loving colleagues' who over many years grasped what I was trying to do and (in Harvey's words) 'were us'd to stand by and assist me' in my researches and teaching while I was a member of the Wellcome Unit for the History of Medicine in the Department of History and Philosophy of Science at Cambridge University. With the support of these colleagues and friends these were golden years for me in teaching, research and writing.

First I dedicate it to the late *Roger French* for the many, many happy hours we spent over 20 years talking about Harvey – our boy. I trust that our two books on Harvey will be found to be complementary.

Then I dedicate it also to *John Gabbay*: in the early days we learnt together how to teach history, partly in the Eagle; to *Adrian Wilson*, for many passionate discussions on how to do history, late into the night; and to *Perry Williams*, especially for a 'big picture' paper we wrote together and which continues to challenge the foundations of our discipline. It was a privilege, a pleasure and an inspiration working with each of you, as it was also with another friend and colleague no longer with us, *Harmke Kamminga*, who very much kept me on my political toes.

# Contents

# Figures

# Preface

Given my current age, mental energy and retirement status, this is likely to be my last book. But it is also, in a way, my first book, one which was conceived in my mind, virtually entire, very early in my career, but which has taken a long time getting itself completed as a whole. This book has been on my mind for many years now, especially the question of how to write it. In one of his *Pensées* Pascal wrote, 'The last thing one knows in constructing a work is what to put first' ('La dernière chose qu'on trouve en faisant un ouvrage, est de savoir celle qu'il faut mettre la première'). And so it has been with this book. Every new start turned out to be a false start. Everything needed to be told at the beginning, and I could not find a coherent sequence for everything that had to be said.

But suddenly, as I struggled with the problem yet once more, inspiration struck. More precisely, a couple of sentences by Harvey, with which I had been familiar for all those years, came into my mind, and *voilà*! all the old plans for the book were quietly put aside, and I could at last begin by writing the title page.

I have recently learnt that the inspiration of Richard Wagner's opera *Parsifal* first occurred to him in 1845 when he was a young man of 32 years of age. Twenty years later the whole work had been formed in his mind. A mere 17 more years passed before he finally realised the work, and it was at last complete and performed in 1882 at Beyreuth, his final opera. I take comfort in the fact that 37 years elapsed between concept and completion in the case of *Parsifal*, for that – and more – has been the case in this present work too. Not that I am comparing them as works of art of course, only as obsessions, though Wagner's *Parsifal* and the work of William Harvey as related in the present book, do have this in common, that they are both overwhelmingly concerned, in their different ways, with blood and with sex. But I am thankful that in the slow building of the present book I did not, like Wagner with *Parsifal*, have to go through the experience of reading Schopenhauer's *The World as Will and as Representation* before I could fully realise it.

Of course in a volume such as this, which is the culmination of many years of work, it is inevitable that I redeploy articles, arguments and texts I have used and published earlier, and for that I make no apology. In a couple of instances I have even used some of my earlier text pretty much wholesale:

I tried hard to re-write these sections, but found I could not improve them after all. However, much of what I have to say is new. I hope no-one feels short-changed.

Readers will notice that I quote Harvey's published writings in their English versions, which appeared shortly after the Latin versions. Most Harvey scholars – including myself for many years – have assumed that the 17th-century English versions were translations from Latin originals. I do not agree, and my reasons are given in the Appendix.

In writing this book I have always had the glorious privilege of working regularly in three of the great libraries in the world: Cambridge University Library, the British Library and the Wellcome Library.

I want to thank two people very close to me for their assistance. I thank David Lindsay Lee for his friendship and his artistic talent, and I thank Yoko Mitsui for continuing to be my first and best reader, and being my bridge between Stone Age technology and the 21st century.

Finally, as my whole academic career was supported in one way or another by the Wellcome Trust, I want to thank the Trustees heartily once more.

# Prologue
# 'Nine years and more'
## An overview of the story

This Prologue is intended to serve as an outline or epitome of my argument about how Harvey discovered the circulation of the blood, with some more heavy scholarly issues around my interpretation being dealt with in subsequent chapters. It is an expanded and lightly referenced version of a lecture I was invited to give at the Royal College of Physicians of London in January 2018, to celebrate the 500th anniversary of the founding of the College. I called it 'How William Harvey discovered the circulation of the blood, and why he regretted it'. My lecture was filmed and made available on YouTube, https://www.youtube.com/?v=ZR8LmpfkXhQ, but as the survival of such new technologies cannot be guaranteed, I put it now into print. I always speak impromptu, with only the PowerPoint slides nowadays as my guide, so there was never a full text that I could reproduce here. This version elaborates several points from that lecture, corrects a couple of errors and leaves out all the jokes.

### Three founders of the College of Physicians

I began to write this in the beginning of 2018, in which year the Royal College of Physicians of London celebrated 500 years since its foundation. And 2018 is also possibly – even probably – the 400th anniversary of William Harvey discovering the circulation of blood. For in the Epistle Dedicatory of his 1628 book announcing the discovery, Harvey wrote that it was being published after 'being confirmed by ocular demonstration for nine years and more' by the President and the Fellows of the College. Nine years before 1628 would give us 1619 as the year of the discovery, whilst 'nine years and more' would give us 1618, or possibly 1617.

That is something well worth celebrating, because this discovery is simply the most important discovery about the functioning of the human and animal body ever made – and it was made by one man working alone and far from the centres of anatomical investigation of the day. The discovery totally underlies all our thinking today about the body and its functioning, it underlies all our interventions to preserve health and save lives, and it underlies all our research on health, disease and drugs, so much so that the extraordinary and revolutionary nature of the discovery is easy for us to overlook. We take

DOI: 10.4324/9781003247616-1

it for granted, and forget that it was once an unthinkable thought and one which sounded like heresy when Harvey first announced it. But more than this, the discovery was not something that was being actively sought, either by Harvey or anyone at all before his time. Nor was it the final outcome, as it were, of a cumulative process of discoveries over many years, as historians have often presented it to have been. Nor was it a welcome discovery, either to Harvey himself or to any other physician. Indeed, Harvey himself, the discoverer, was as shocked by the discovery as anyone and for many years he was to regret having made it and having published his findings.

As will become evident, I regard the London College of Physicians, modest institution as it then was, as being crucial to Harvey's discovery and announcement of the circulation of the blood. It is not just a matter of Harvey being a Fellow of the College when he made his discovery, but of Harvey succeeding in getting the College to play a critical supportive part in his work.

I want to start this story of the discovery of the circulation of the blood, linked as it is with the London College of Physicians, as a tale of two cities: London, the capital of England, a growing capital city without a university, and Padua, the university town of the Republic of Venice.

The first founder of the London College of Physicians was Thomas Linacre (c. 1460–1524) (Figure P.1), an Englishman from Canterbury.[1] After several years spent at Oxford University, he travelled for a time in Italy. Whilst there he went to Padua to take the medical degree of M.D. which he was awarded in 1496. He returned to London where he practiced and became a royal physician.

*Figure P.1* Thomas Linacre, first founder of the College of Physicians. (Wellcome Images).

In time he obtained a charter in 1518 from the king, Henry VIII, to found a college to control and improve medical practice in London. The wording of the charter (in the king's voice) said that in founding this college, 'we have copied the example of well-run cities in Italy and in many other nations'.[2] So the institution of a College of Physicians is something that Linacre imported from Padua and other cities in Italy to improve the state of the practice of medicine back in London, and to help bring London medical practice and institutions up to Italian standards. That meant medical practice based on education in Latin, and Greek too if possible, built, above all, on the newly edited and translated medical writings of the two great Ancients of Greek medicine, Hippocrates and Galen. The role of the College in London was at first to be a *policing and licensing* institution to defend and promote the best practice of medicine.

But this was not all. For Linacre had himself learned newly fashionable Greek and when he was in Italy he could, and did, take the opportunity to join in the movement of translating ancient Greek medical texts into Latin to make them more available to his contemporaries, including texts written by the great ancient Greek authority on medicine, Galen.[3] All this indicates that Linacre was fully integrated into the Renaissance ideal of renewing medicine by resort to perfect Latin translations of the Greek works of Galen. England was not just on the physical outskirts of Europe at this time, but also on its intellectual outskirts, and its two universities were of no account compared to those of Italy. Here Linacre also resolved to improve medical teaching in both Oxford and Cambridge Universities by the founding of lectureships in his name in the 1520s. There were to be two of these, one in St John's College in Cambridge, the other in Merton College, Oxford. Both lecturers were expected to teach from the works of Galen, especially in Linacre's Latin translations. In the event, however, both of these lectureships had only limited success in establishing the new kind of medical teaching in Britain.

Any institution has to move with the times if it is to survive, and this may involve the adaptation of its original aims, or extensions of them. This, I suggest, is what we see in the case of the London College of Physicians. So when John Caius (Figure P.2), the man celebrated as the 'second founder' of the College went, in 1533, to Padua for his medical education (like Linacre had done almost 40 years earlier), he also engaged very actively in the recovery of the medical writings of Galen and others. He was in Padua at the time of Vesalius's dramatic revival of anatomy, and when he came back to London he, too, taught anatomy: he later recalled 'the anatomical dissections which we gave for surgeons in London at the instigation of that illustrious prince Henry, the eighth king of that name. (There were none at the time for physicians)'. These dissections were overseen by Caius for about 20 years.[4] As well as introducing the regular dissection of dead human bodies for teaching, Caius also raised the status of the College by writing the history of the College and making its affairs grander and more ceremonial, as befitted the elite status of learned physicians.

Like Linacre before him, Caius also turned his updating reforms to one of the two English universities. In the first place he re-founded Gonville Hall at

*Figure P.2* John Caius, second founder of the College of Physicians. (Wellcome Images).

Cambridge in 1557, as the College of Gonville and Caius. Within this renewed college, Caius provided for two medical scholarships and acquired a grant from the Crown that two bodies of criminals put to death should be provided to be dissected each year in the presence of medical students. As with Linacre's attempts to modernise medical teaching in Oxford and Cambridge of 30 years before, so Caius' attempt to provide bodies for dissection was very innovative – one might say very Italian – though it had limited success.

I want now to suggest that we could with justice see William Harvey (Figure P.3) as the third founder or perhaps re-founder of the London College of Physicians, in that he too went to Padua for his medical education and when he returned he proposed and put into effect new goals for the College and its members. Not only did he himself engage in anatomical research as he had learned it in Padua, but he also encouraged the Fellows of the London College to engage in research as well, like in Padua. In other words, he tried to add the function of *research institution* to the existing functions and goals of the College as established by Linacre and Caius.

William Harvey was very much a College man, indeed we could probably regard it as his second home. Harvey was a regular attender at the meetings of the College and held several offices, though he declined the position of President. He speaks of his 'most loving colleagues' there, and he recruited them time and again over several years to witness and test his strange and unexpected discovery.

Indeed, for many years, from 1615 to 1656, Harvey took on one of the most important roles in the College: giving the annual anatomical lectures on the human body. Presented primarily for surgeons, these lectures, named after the original donor, Lord Lumley, had begun in 1584 and went on a

*Figure P.3* William Harvey, discoverer of the circulation of the blood. Portrait attributed to Daniel Mytens. (National Portrait Gallery, London).

six-year cycle. Ideally a whole human body was dissected in the first year over a period of up to five days, and different sections of a human body in each of the other five years of the cycle. Carrying out this duty would have made Harvey skilled in anatomy and dissection, far more skilled and knowledgeable than anyone else in England.

At his own expense and whilst he was still an active member Harvey donated a new building to the College as a library and depository for rarities and examples of the basic ingredients of medicines in 1651–1652. Unfortunately all this got destroyed in the Great Fire of London of 1666. But fortunately many of Harvey's extensive manuscripts escaped this fate. The notes from which he gave his annual anatomy lectures at the College still survive, as do several other anatomical manuscripts, for in time they were to find their way into the collections of Sir Hans Sloane, and they are today in the British Library.

Moreover, in his will Harvey left money for an annual feast at the College. The point of this feasting, which is continued to this day, was to bring the Fellows together as in a common enterprise, and to stop them fighting over patients or arguing publicly with one another. As every Fellow was a private practitioner seeking patients, such disagreements frequently broke out. Come together, says Harvey, keep at peace with one another with this shared goal of protecting the proper practice of medicine in London. In his will Harvey also gave 'an exhortation to the Fellows of the College to search and study out the secrets of Nature by way of experiment'. It is this above all which indicates that Harvey wished to develop the role of the London College of Physicians and take it in a new direction, following practices that he had himself learned when he was a student in Padua. So, in addition to its historic role as a defender of good medical practice in London (Linacre) and an institution promoting the most up-to-date anatomical and medical teaching (Caius), in Harvey's view the College should also now be a research institution. This was a new vision of the roles of a non-university medical institution.

In his own research Harvey led the way and was a model to his colleagues, and for a period it looks as though Harvey was successful in encouraging the Fellows to engage in novel and innovative research. In a book published in 1657 Dr Walter Charleton, though not yet himself a member, calls the College '*Solomon's House* in reality'. It was Francis Bacon who, in 1627, had described a fictional Solomon's House, a house designed for the getting and the deploying of new wisdom and knowledge about nature. This, writes Charleton, you can actually see now in the College of Physicians, where 'some constantly employ themselves in dissecting Animals of all kinds, as well living as dead', whilst 'there are those who daily investigate arguments to confirm and advance that incomparable invention of Doctor Harvey, the Circulation of the Blood'.[5]

Several books were published by Fellows mid-century, recounting their research at or in association with the College, and conducted in a Harveian spirit. Francis Glisson, Fellow of the College and also Professor of Medicine at Cambridge University, produced a book which was a collaborative effort by three men, all Fellows of the College (Glisson himself, George Bate and Assuerus Regimorter), on what seemed to be a new disease, the rickets. The authors recalled: 'Before the space of five years, we have mutually communicated by written Papers something concerning this Affect [= disease] in private meetings (which some of us Physitians used sometimes to have for Exercise-sake in the works of Art [viz., medicine])', and it appeared in 1650.

Another work by Glisson, which appeared in 1654, is on the anatomy of the liver – the first such detailed anatomical investigation of a single organ – which resulted from lectures Glisson had given at the College in 1641. As Glisson recalled, 'according to the custom of the College at that time, they were read in English', but other Fellows thought that before they were published they 'should first be turned into Latin, the language of the learned'.

And yet another book was produced by another College Fellow as the outcome of his own 'Harveian' researches: Thomas Wharton wrote *Adenographia*, which is on all the glands of the human body. Published in 1656 this, like Glisson's book on the liver, records Wharton's findings from his investigation of a single type of organ and again was similarly the result of lectures he had given in the London College of Physicians in 1652. He dedicated 'this anatomical exercise (*exercitatio*)' to the College and to his four best friends there, Baldwin Hamey, Francis Glisson, John Bathurst and George Ent.[6]

Outside of the College there were other medically trained men who engaged in what was informally called a college of anatomy, mostly based in Oxford, and often invoking the name of Harvey. Such were Dr Thomas Willis, who was at the centre of a little group of investigators, and who produced *Cerebri anatome*, in 1664 (*The Anatomy of the Brain*). Helping him there was Richard Lower, who in turn produced his *Tractatus de corde* in 1669 (*A Treatise on the Heart*). In that group was also Walter Needham, who published *De formato foetu*, in 1667 (*On the Formed Fetus*). All of them had connections with the London College of Physicians: Willis was made an honorary Fellow in 1664, Lower became a Fellow in 1675, and Needham was admitted to the College in 1664. This is quite a crop of original investigative writings recording work carried out in the wake of Harvey's own experimental work by men who were Fellows of the College before, during or after their investigations.

It will be noticed that all these books are in Latin even though some, such as Glisson's book on the liver, were originally written in English. It is because their authors were seeking a Europe-wide audience that these books were produced in Latin, since English hardly counted as a scholarly language at the time. So by following Harvey's exhortation, English investigative physicians began to have a voice in the greater scholarly arena of Europe. This was quite a change from the days when Harvey returned to London from Padua, when according to the gossipy writer John Aubrey in his *Brief Lives*, written in the 1680s, Harvey was 'the first that I hear of that was curious in Anatomy in England'.

## Blood movement before Harvey

Harvey established that the blood of animals, including man, is pumped around the body fast and continually, being pushed out from the heart into the arteries, pushing on into the smaller and smaller arteries, and then being taken up into the smallest veins, and passed on to successively larger veins, until it returns to the heart. From which it is again pumped out into the arteries, right throughout the life of the animal, and when the heart stops pumping the blood round the body then the animal dies. It is a single blood system, with the same blood being passed out into the arteries and then returning to the heart via the veins. One blood, one system of flow, in one direction, round and round. This is what we all learn at school, in life, in hospital, about the movement of the blood. All our understanding and practice of medicine is based upon it. Every daily accident or intervention we make, such as cutting ourselves, tending grazes, staunching blood flow, giving blood voluntarily, measuring blood pressure, administering injections, all depends on and confirms this common-sense view of blood flow. It has become so obvious to us that we hardly remark on it. Every educated and uneducated adult in western Europe has been taught that this is the case, and every encounter he or she has with the blood appears to confirm it.

But the understanding of blood flow in animals, including man, *before* Harvey's discovery was totally different, yet equally common-sensical. And this understanding in its turn was also shared by every educated and uneducated adult in western Europe. Similarly, every encounter he or she had with the blood appeared to confirm it. And this was not some folksy view, this view of blood flow in humans: it came from the highest level, it came from

2,000 years of anatomists performing dissections and undertaking research, and from highly educated physicians in their medical practice.

The easiest route to understand how this was the case will be by a whistle-stop history of the practice of anatomising – brief, I promise.

We start here with Aristotle in the 4th century before Christ, living, teaching and working in Athens. This is the same Aristotle who was a pupil of Plato and wrote so many books on philosophy. He is the earliest person we know who systematically dissected and vivisected animals to find out how and why they work, which is what anatomising is. He wrote several books on animals, built on knowledge gained from anatomising by himself and his contemporaries and predecessors, and these books survive today. Anatomising was always essentially a Greek enterprise: classical Greek culture is the only one in which such systematic investigation took place.

After the death of Aristotle, for 500 years Greek-speaking philosophers and physicians continued to dissect bodies and to discuss the workings of the body, especially the human body. But their writings, unlike those of Aristotle, almost completely disappeared.

But we know something of what they did and wrote from our major classical sources on anatomy, the writings of Galen. Working in Rome in the second century after Christ, Galen was a Greek physician and anatomist, writing in Greek. He actively dissected and vivisected animals primarily to find out how the human body functions. But even though he was physician to the gladiators in Rome, Galen was forbidden to dissect humans after their death to investigate their structure and workings. It seems perverse to us that the Roman public could enjoy battles to the death between gladiators in the amphitheatres of Roman towns, and between gladiators and wild animals, yet a conscientious medical investigator such as Galen was forbidden, for cultural reasons, to dissect them after death. But such was the case. So Galen's knowledge of human anatomy comes most directly not from humans but from animals, especially the ape. And Galen learned a great deal in this way and wrote about it at length. Not only that, but Galen's enormous writings have survived almost complete. Naturally enough his works on anatomy and medicine later came to be viewed as authoritative in themselves and as trustworthy summations of all the lost work that had been done by previous investigators.

The works of Galen, only partially available for hundreds of years, all became available in Western Europe in the Renaissance, that movement to give 're-birth' to Greek and Roman learning, and to Greek and Roman skills and practices, a period running roughly from 1350 to 1600. The centre of Renaissance activities was Italy, and it is here that most of the ancient Greek texts on all sorts of subjects were recovered and translated into Latin.

This was also the case with the anatomical writings of Galen, and the culminating moment here was the publication in 1543 of the *Fabrica* as it is called for short, the enormous volume entitled *Seven Books on the Fabric of the Human Body, by Andreas Vesalius of Brussels, Professor of the Paduan School of Physicians*. This was Vesalius emulating, in a typical Renaissance

*Figure P.4* Title page of the *Fabrica*, 1543, by Vesalius (a) Detail of the title page of the *Fabrica*, 1543, by Vesalius. (Author's collection).

way, the anatomical achievements of Galen and – because he was also an arrogant young man – also criticising Galen loudly, whenever he thought it necessary. The title page (Figure P.4), a masterpiece of wood-block cutting, shows Vesalius himself conducting the dissection of a human. The title is about the fabric of the *human* body (Figure P.4a).

Vesalius was not a shy young man, as we can see from this portrait of himself which he included, saying that he was only 28 years old (Figure P.5).

*Figure P.5* Self-portrait of Vesalius in the *Fabrica*, 1543. (Author's collection).

This *Fabrica* was not just a book of text, as Galen's anatomical writings had been, but Vesalius also put the whole of Galen's understanding of human anatomy into fantastic pictures. There are hundreds and hundreds of exquisite artistic wood-cuts in the book, which made it – and still makes it – far and away the most stunning and original book on human anatomy ever published.

Here Vesalius was exploiting the still fairly recent inventions of printing and the wood-cut. Galen had not used illustrations in his manuscript writings because he knew that as they were copied out by unskilled copiers over the years, so they become corrupted and inaccurate and worse than useless. But now with printing, an image can be, as it were, 'frozen', and produced in hundreds of identical copies.

But before he even thought of writing the massive *Fabrica*, Vesalius had issued what are now called the *Six Tables* in 1538–1539. These are totally original in concept and in execution. They are six separate large, printed sheets ('tables') portraying aspects of the anatomy of the human body. They were probably drawn by Vesalius himself and cut as woodblocks and printed in Venice. In virtually every detail they illustrate what Galen and later anatomists thought.

The *first* of these tables (Figure P.6) shows the source of the blood. Veins run from the wall of the intestine to the liver. They carry ingested food and in

*Figure P.6* Vesalius: Table 1 of the *Tabulae Sex*: on 'The liver, the workshop of blood-making'. (Wellcome Images, Stirling-Maxwell facsimile).

the liver this is made into blood. So, the source and origin of the blood is the liver.

The *second* of these tables (Figure P.7) is on the *vena cava*, the great hollow vein running vertically through the centre of the body, which is portrayed as being based on the liver and which distributes this blood to all the parts of the body through the veins. You will notice that the liver is the only organ portrayed. These are major blood vessels of the body, but what is not present? The heart!

The reason for the absence of the heart from the portrayal of the veins is that to Vesalius – just as to Galen and everyone after him – it is just not relevant. For the heart is, instead, the centre of the *other* set of blood vessels, the arteries, as we see in the *third* table (Figure P.8). Whilst Vesalius does not portray the lungs in the *Six Tables*, he does portray them later in the *Fabrica*, as is seen in the next image (Figure P.9), where you can see the trachea (or windpipe) running down from the mouth to the lungs, and then the lungs almost enveloping the heart. For, to the whole Galen-Vesalius school of anatomists – which is virtually everyone – the heart is an organ of *respiration*, conveying some vital life ingredient from the air, via the lungs, to the arteries. Let us call this vital something 'spirit'.

We can see from this that the anatomists believed there were *two virtually separate* blood systems in the human body, not one. Both of them are needed if you are to stay alive: one nourishes you (the blood in the veins) and one vitalises you (the blood in the arteries). We can also see from this dash

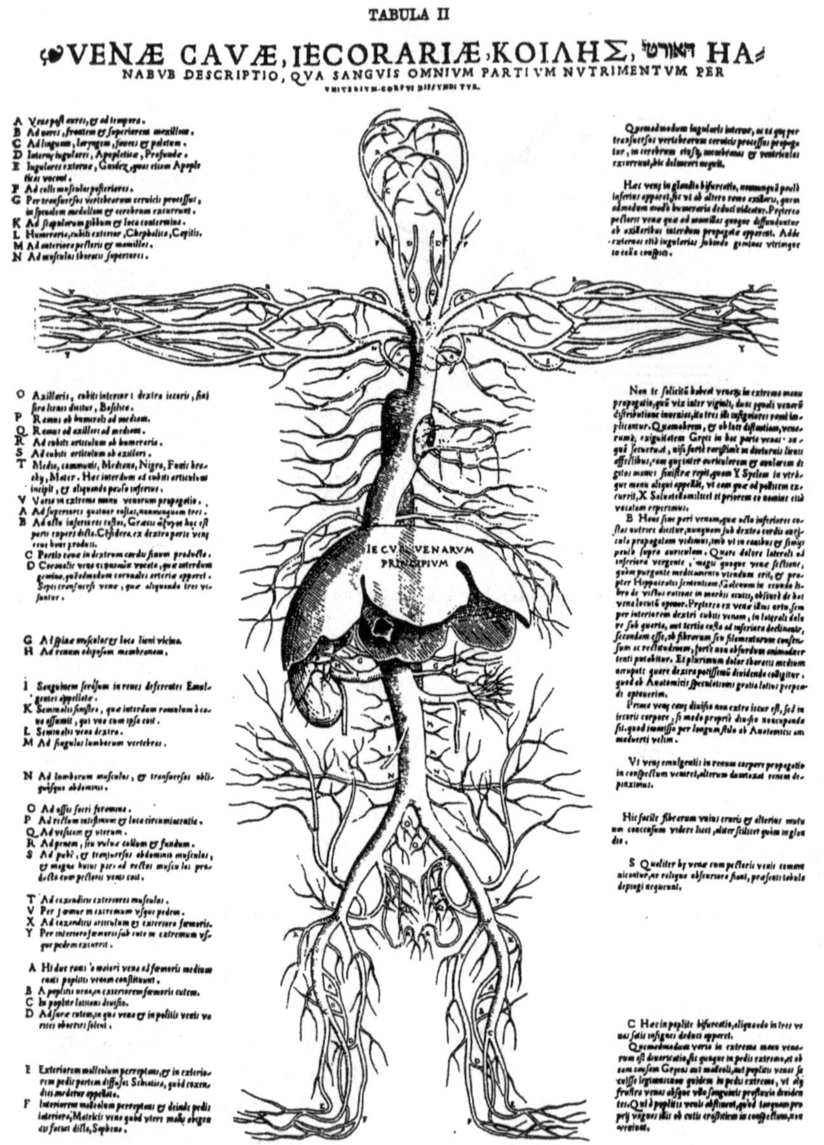

Figure P.7 Vesalius: Table 2 of the *Tabulae Sex*: 'Description of the *Vena Cava*, by which the blood, the nutriment of all the parts, is spread through the whole body'. (Wellcome Images, Stirling-Maxwell facsimile).

through the history of anatomising, that the Galen-Vesalius – that is the pre-Harvey – understanding of blood flow in the human body was very common-sensical, and had been confirmed by the research of anatomists over many centuries.

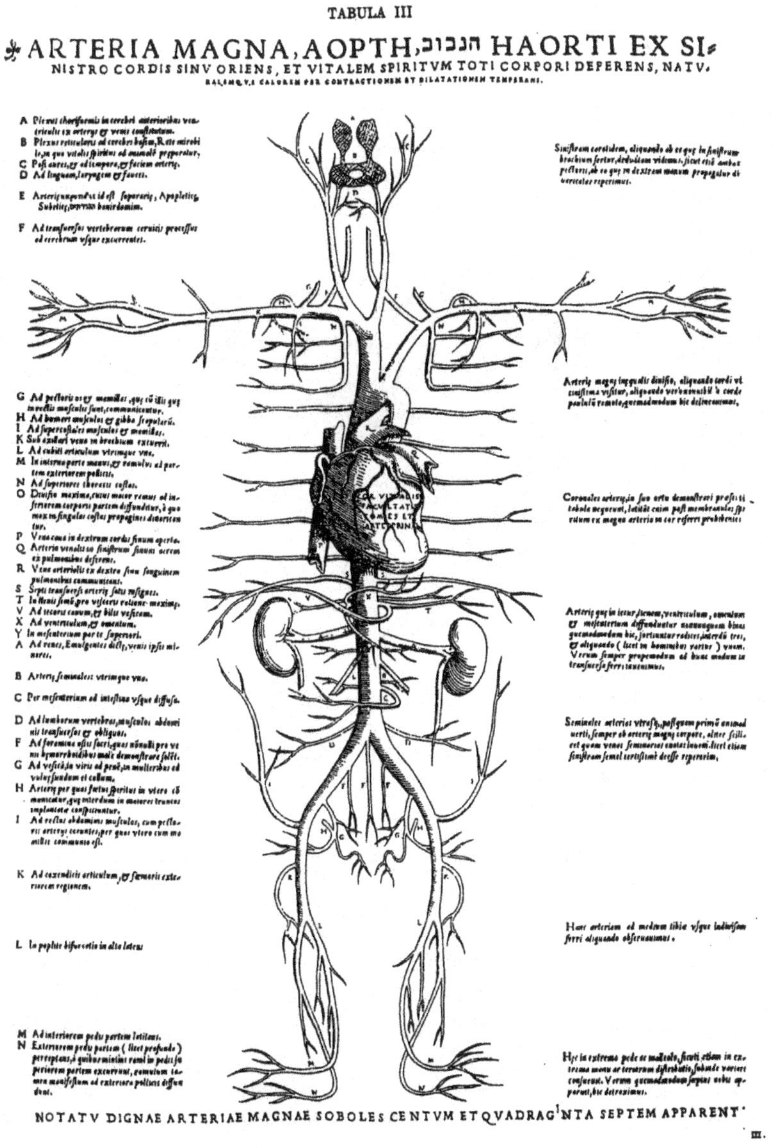

*Figure P.8* Vesalius: Table 3 of the *Tabulae Sex*: 'The *Great Artery*, arising from the left cavity of the heart, both bringing the vital spirit to the whole body, and tempering the natural heat through its contraction and dilation'. (Wellcome Images, Stirling-Maxwell facsimile).

Everyone, including the man in the street, knew that if you cut an artery you will die quickly unless you can staunch the flow urgently. The blood spurts out. But in normal life arteries are quite hard to find, as they lie deep in the body and limbs, and they are therefore quite hard to cut accidentally. They come to the surface (as it were) most noticeably at narrow points of the human body

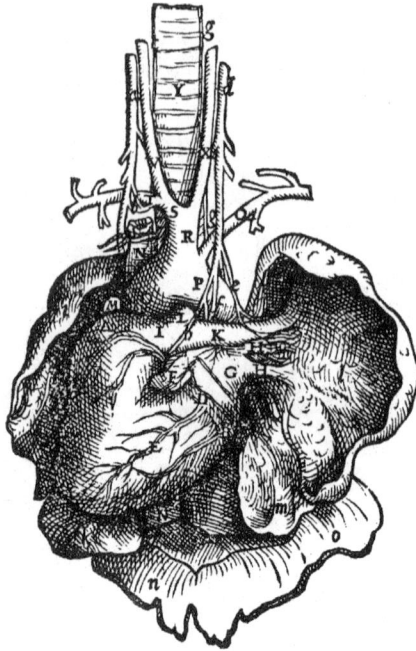

*Figure P.9* Vesalius: One of Vesalius's figures showing the lungs and heart of the human; note how the lungs are shown as surrounding the heart, both involved in respiration. The sixth figure of the sixth book of the *Fabrica*, 1543, p. 564. (Author's collection).

– in the wrist, the ankle and the sides of the neck – and in these places we can put our finger on the actual artery beneath the skin and feel the pulsation of the blood within it. But if, by contrast, you cut a vein, which is far more likely as they are far more numerous and nearer the skin than arteries, then you are unlikely to die quickly because in normal circumstances the blood will coagulate or clot and the flow will soon stop. And unlike the arteries, the veins do not pulsate, nor does the blood spurt out of the veins as it does out of the arteries.

So much for the views of the man in the street before Harvey's discovery. But if you were an anatomist in the medical faculty of a university or in a professional medical college and were investigating the veins and arteries in a dead human or animal body, you could see things more closely. And then it was quite obvious to sight and touch that the two kinds of blood vessel differed not just in pulsation when alive (the arteries pulsate, the veins do not) and location (the arteries deep in the body, many veins often accessible at skin level), but also in structure. For you would find that the walls of the arteries are much thicker and stronger than those of the veins. And if you were an experienced surgeon or physician you could also see a difference in the colour and liveliness of blood springing out from the pulsating arteries when cut, and in the blood flowing from the non-pulsating veins of a living creature. You would see that the blood from the arteries was brighter red and more spirituous and lively, whilst that from the veins was darker and more sluggish. So everything that the physician or surgeon did to the body and which involved blood confirmed the fact that there were essential and physical differences between the blood in the arteries and the blood in the veins, and also confirmed the common-sense view that they were two different kinds of blood in two different systems of blood flow, located in two distinct and different structures, for two different purposes in the life of man and other animals.

There were no problems with any of this, whether to the sight or the touch or in medical practice, which might induce someone to wonder whether there really are two systems of the blood, or might lead one to suspect they were just one system. It all fits, and it is all common-sense. And, again, no-one was looking for inconsistencies in the Galenic account of blood flow: why should they?[7]

## Back to Harvey

How did William Harvey discover the circulation of the blood? Unfortunately for us the book announcing the discovery, *De motu cordis et sanguinis in animalibus, On the Motion of the Heart and Blood in Animals,* is not a laboratory notebook in which we could trace the sequence of the discovery. It is, rather, an argument. We can search for clues about the stages of the discovery by using Harvey's handwritten text of the lectures on anatomy that he gave over many years in the College. But even here the manuscript changes that Harvey made over time in these lectures are hard to date precisely.

But if a search of Harvey's texts in this way is not very helpful, we can find a large part of the answer in his training. For Harvey goes to Padua! And it is in Padua that Harvey becomes fascinated by the possibilities of new anatomical investigation, and also gets his new 'problematic' in anatomy, that is, a new way of setting up problems or topics to investigate.

And when he arrives in Padua in early 1600, there are new and improved facilities there since the time of the visits of Linacre and Caius, all of them keeping Padua at the forefront of universities in Europe and making it the most exciting and attractive place for a bright and ambitious young scholar or medical student.

First there is the physic or botanic garden (Figure P.10). The one at Padua was the first in the world, established in 1545.[8] It is still there and can be visited. Similar gardens were soon founded in other universities.

Second, there are the new university buildings, completed in the 1590s, named after an old inn that stood on the site, Il Bo (The Ox). These can still be seen today, as they are still in use, especially for ceremonial purposes, though the students no longer feel obliged to carry swords (Figure P.11).

If we take the front of the building off, as is done in the next image, we can see the third and perhaps most important structural innovation here at Padua: the world's first permanent anatomy theatre. It is up at the top left corner (Figure P.12).

This theatre was of a very ingenious design, and there were standing positions for up to 300 people! It continued to be used for teaching until 1872 (Figure P.13). It is still there and you can visit it today.

*Figure P.10*  View of the Padua Botanic Garden today. (Photo by Yoko Mitsui).

*Figure P.11*  Gymnasium Patavinum, Palazzo Il Bo. Exterior view of Padua University, 1590s. (Wellcome Images).

THEATRVM ANATOMICVM

PARS INTERIOR GYMNASII PATAVINI

*Figure P.12* Gymnasium Patavinum, Palazzo Il Bo. Interior view of Padua University, 1590s. (Wellcome Images).

It was in the anatomical theatre that young Harvey first encountered the Professor of Anatomy, Girolamo Fabrizi d'Aquapendente, whom we nowadays refer to as Fabricius, the Latin version of his name (Figure P.14). He had been Professor for over 30 years when Harvey arrived, and he was in the course of writing up his anatomical researches over those years in a work called *The Theatre of the Whole Animal Fabric*, which began appearing in parts in 1603.

The anatomical research of Fabricius was modelled after that of Aristotle, rather than Galen or his modern followers. This may seem surprising at first sight, but in fact Fabricius' alignment with Aristotle in anatomising was of a piece with his eminent colleagues in the philosophy faculty at Padua. One could say that both the philosophers and the anatomist had gone back to the first, basic principles. And this meant in both cases that they were doing something essentially new, whilst also essentially ancient.

Fabricius' *Theatre of the Whole Animal Fabric*: whilst everything he discusses is relevant to the anatomy and working of the human body, it is not centrally about the human body but – as the title says – about the *animal* body. By following Aristotle, Fabricius has new projects and new approaches. Here is an image from his booklet on vision (*De Visione*), which explores the organ which provides vision, that is to say the eye, in its differing incidence in different animals (Figure P.15).

In this anatomical theatre young William Harvey became completely absorbed and fascinated by this Aristotelian approach to anatomy and adopted it wholesale as a lifetime project.

*Figure P.13* Padua Il Bo anatomy theatre in the mid-19th Century. From Pietro Tosoni, *Della Anatomia degli Antichi e della Scuola Anatomica Padovana Memoria*, Padua 1844. (Author's collection).

*Figure P.14* Portrait of Hieronymus Fabricius, Professor. (Wellcome Images).

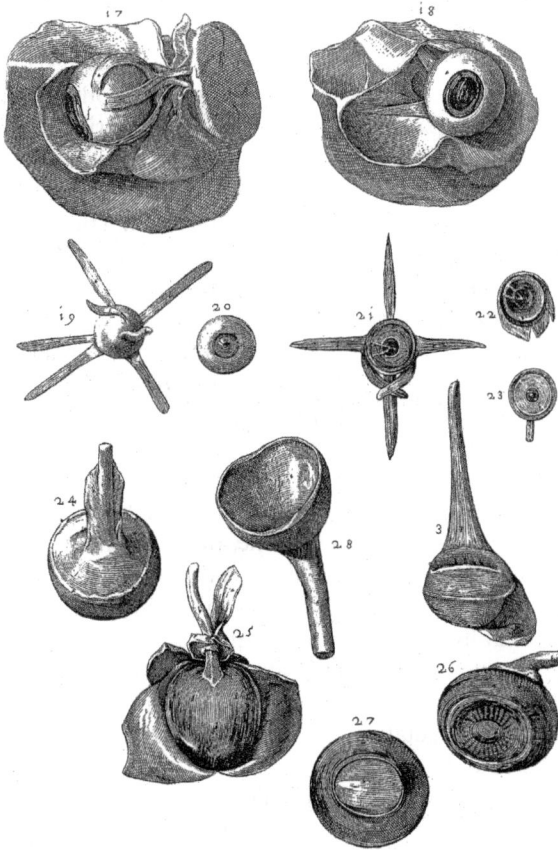

*Figure P.15* Eyes of different animals from Fabricius's *De Visione*, 1600. (Wellcome Images).

Harvey returned to London in 1602 or early 1603 with his Padua M.D. He began to earn his living as a private physician, as a Royal Physician to James I and then to Charles I, and as Physician to St Bartholomew's Hospital. He was also the anatomical lecturer to the College of Physicians for many years.

On top of all this he privately performed anatomical research on live and dead animals and on dead humans. He was a passionate dissector of every kind of animal he could get his hands on. The list of animals that Harvey dissected and vivisected is very long. He himself lists 'hens, geese, pigeons, ducks, fishes, shell-fish, molluscs, frogs, snakes, bees, wasps, butterflies, silk-worms, sheep, goats, dogs, cats and cattle'. He also dissected 'the most perfect of all creatures, man himself'. And even his wife's parrot when it died.

Historians still argue over whether Harvey's work is Aristotelian or modern. His book on the motion of the heart and blood in animals hardly mentions Aristotle. Does it therefore fit in with the 17th-century movement which we historians call the 'Scientific Revolution', with its promotion of mechanics, mathematics, measurement and its rejection of the physics of Aristotle? Yes, say Harvey's modern biographers.

But I say no. Harvey repeatedly puts his own research into the tradition he has learned at Padua, that of Fabricius following Aristotle. Harvey says in his famous book, 'I am persuaded to publish because Fabricius ... having learnedly and accurately set down in a particular treatise, almost all the parts of living creatures, left only the *heart* untouched'. And if we look at the title page of that book (first published in Latin) announcing the discovery in 1628, it clearly states that this is *An Anatomical Exercise on the Motion of the Heart and Blood in Animals* (Figure P.16). Compare this with the title of Vesalius' book of 1543: *Seven books on the fabric of the **human** body*.

Moreover, Harvey's other writings are explicitly Aristotelian, especially the *Anatomical Exercitations Concerning the Generation of Living Creatures* first published in Latin in 1651. There is an extensive introduction to this book on what method to follow when investigating nature, and it is very, very Aristotelian. 'I follow Aristotle ... as my General, Fabricius ... as my Guide', he writes about his approach in anatomising. A friend of his, one John Twysden, writing in 1666, says, 'Aristotle, of whom no man was a greater admirer than Dr Harvey, who has often said that he was the most rational and acute Philosopher that every lived, that his writings were near divine ...'.

It should be clear by now that Harvey will be an Aristotelian in his research practice. He will study 'the animal', 'all animals', 'living creatures', not just man. He will be seeking a universal understanding of the actions of the soul – especially the vegetative soul, which controls or directs all the inner workings of the body – through the organs.

So when he begins his private anatomical research programme in London, Harvey naturally enough follows the programmes of Aristotle and of Fabricius. But he also starts a quite new one. He mentions it in connection with why he is publishing on the heart: because Fabricius had left the heart untouched. This is the crucial project, the one which will lead Harvey, all unwittingly and unwillingly, to the discovery of the circulation of the blood.

When it comes to the heart, Harvey wants an account of why 'the heart' is how it is, what it does, what it is for. As hearts in different animals differ in shape and structure, he wants an account of the anatomy of the differing hearts, so that from these he can work out what 'the heart', every heart, does as a heart. And then he can work out (following Aristotle) why the hearts of different animals differ, according to (1) their life, (2) their activities, (3) their habits and (4) their other parts. He wants to understand the 'final cause' of everything – for what purpose?

It is an investigation of the motion of the heart and blood – the heart and blood of all animals – without looking at respiration. And in the human that means not looking at the lung. He has a new object of research, plus a new/old approach, that of Aristotle. I cannot stress enough that Harvey was the first person ever to draw up the research topic quite like this. Why have other investigators, Harvey asks, not seen the route of the blood? 'Since it is probable, that the connexion of the *heart* with the *lungs* has given this occasion of mistake,

## EXERCITATIO
# ANATOMICA DE
## MOTV CORDIS ET SAN-
### GVINIS IN ANIMALI-
#### BVS,

### GVILIELMI HARVEI ANGLI,
*Medici Regii, & Profefforis Anatomiæ in Col-*
*legio Medicorum Londinenfi.*

### FRANCOFVRTI,
Sumptibus GVILIELMI FITZERI.
___
*ANNO M. DC. XXVIII.*

*Figure P.16* Title page of Harvey, *De Motu Cordis et Sanguinis in Animalibus*, 1628. (Wellcome Images).

they are to be blamed in this, who ... look but into man only, and into him being dead too ...' (Ch. 6, pp. 26–27). The correct way to proceed in this inquiry is to ignore the means of respiration (the lungs in man), and to look not just at man, but at all animals, and also to use vivisection, the anatomising of the living creature, to see the movement of the heart and the blood in action.

Or, in Harvey's own words:

> Therefore it will be profitable to search more deeply into the businesse, and to contemplate the motions of the *arteries* and *heart*, not only in man, but also in all other creatures that have a *heart*; as likewise by the frequent dissection of living things, and by much ocular testimony to discern and search the truth.
>
> (Proeme)

There was a moment of realisation, probably in 1618, that Harvey recalls, in a long, long sentence. 'Truly when I had often and seriously considered with my self … how great abundance of blood was passed through [the heart], and in how short a time that transmission was done …[and many other things] … I began to bethink my self if it might not have a *circular motion*, which afterwards I found true …'(Ch. 8, pp. 44–45).

With this realisation Harvey has turned the two blood systems of the Galenists – of all educated physicians – into one! This is medical heresy!

My first line of approach here to try and answer the question how Harvey discovered the circulation of the blood, was to look at his education and his inspiration in anatomical research. In other words, in Padua, with Professor Fabricius who was re-animating the work of Aristotle on animals.

My second line of approach is to look at what was one of the turning points of his own investigations, by his own admission to young Robert Boyle. Viz. *folding doors*.

You can see two-leaf doors still today all over Sicily, as in *Montalbano* films: no-one can walk straight through them, they are always allowing, but at the same time hindering, access.

The thing about folding doors, as we see today in our own homes, is that they have at least two leaves, you can go out and go in, but they restrict access to some extent, though they only stop access when they are closed. Folding doors are not new: they have been found in Pompei, surviving the volcano. The next image (Figure P.17) is an artist's reconstruction of a folding door from Pompeii, which would have been covered with ash from the volcano Vesuvius in 79 AD.

Now the word *valvae* in Latin means 'folding door'. It is a plural. You can't have a singular folding door, that would just be a door. Now, in Harvey's day the word doesn't yet mean valve in the modern sense, a structure allowing only one-way flow. Harvey's work here is a major source of the modern meaning. For one of the things Harvey establishes is the competence of these little doors, these little flaps. As recently as 1615 we can find Helkiah Crooke talking in English in a medical treatise of the 'valves which *hinder* the reflux of the choler' – hinder, not stop. One of the experiments Harvey does is take the arm of a human corpse and try to push a probe down the veins (Ch. 13, p. 76). But it is stopped by these little doors, because they only allow the flow of blood in the opposite direction, and they are very robust. The young Robert Boyle asked

*Figure P.17* An artist's reconstruction of folding doors in Pompeii before the eruption of Vesuvius. From T.H. Dyer, *Pompeii – Its History, Buildings and Antiquities*, 1867. (Alamy image).

the elderly William Harvey, 'Sir, what was the moment when you realised the blood must circulate?' Harvey replied, 'When I was investigating the valves in the veins' (author reconstruction of the conversation). In the only illustration in the book, Harvey shows how you can see the presence and the action of the valves in the human arm for yourself – you don't have to perform a dissection on anyone. You can stop the blood between these little nodules with your fingers. The blood only goes one way. This is actually an illustration Harvey took from Fabricius – where Fabricius thought the little doors were delaying the outward flow of the blood in the veins! (See Figure P.18.)

So, yes, these *valvae* are indeed 'little doors', but doors which only allow one-way flow of blood in the veins.

Harvey has discovered this awful truth – the circulation of the blood – which contradicts what everyone has thought for 2,000 years. The text of the book, I imagine, lay on his desk, burning a hole in it, as it were, for nine or more years, whilst Harvey continued to urge his 'most loving colleagues' to 'stand by and assist me' in making sense of the discovery and encouraging him to publish.

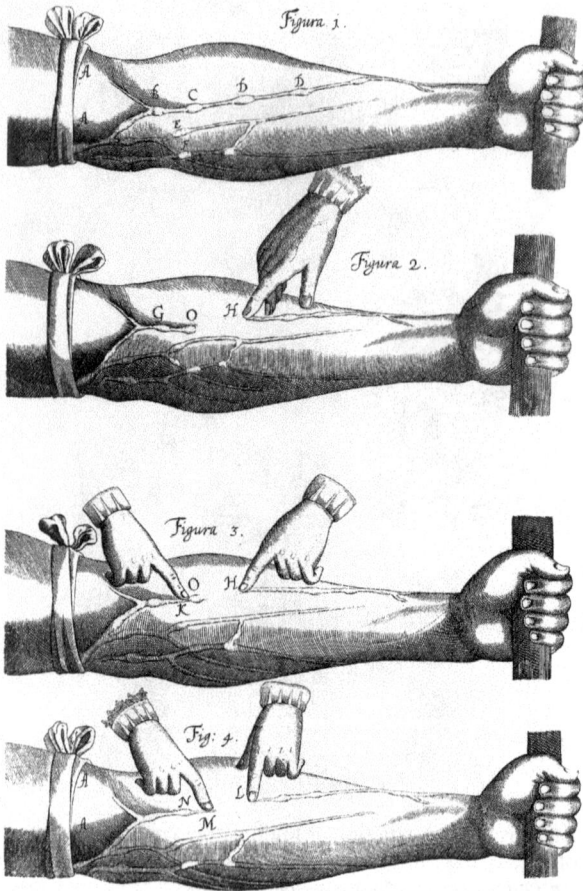

*Figure P.18* How to see the valves in the veins. From Harvey, *De Motu Cordis et Sanguinis in Animalibus*, 1628. (Wellcome Images).

Harvey was in no way a revolutionary, wishing to stir up medical thinking. He was a conservative in politics, a physician to two kings, and even went to the battlefield with Charles I. He was a conservative in medical politics too, continuing firmly in the College of Physicians, even when everything it stood for was under attack during the Commonwealth period.

Moreover, Harvey hates the new approaches being put forward in physiology and medicine in his day. The diarist John Aubrey sought his advice when he was on his way to Italy, briefly wishing to become a doctor in Padua. 'He dictated to me, what company to keep, what books to read, how to manage my studies: in short, he bid me to go to the fountain head and read Aristotle, Cicero, Avicenna, and did call the neoteriques shitt-breeches.'

Harvey also had no time for people advocating chemical understandings of the workings of the body. And in particular he had no time for Descartes and his atomistic explanations of physiological phenomena. Descartes co-opted Harvey's discovery into his theory of the heart functioning as a heat engine, which did not please Harvey.

Harvey had regrets. We have seen how he was hesitant to let the world know about his unwelcome discovery. He feared the worst. He wrote: 'which may arrive to me from the envy of some persons, but I likewise doubt that every man almost will be my enemy'. And he was right: the discovery met with widespread attack. But also more than this. John Aubrey again:

> I have heard him say that after his Book of the Circulation of the Blood came out, that he fell mightily in his Practice, and that 'twas believed by the vulgar that he was crack-brained. And all the physicians were against his opinion.

We can see that Harvey tried to take precautions. In the first place he deferred publication – for nine years and more. He also sought the support and eye-witness of his colleagues at the College, and of the President. 'I first propounded it to you, and confirmed it with ocular testimony, answered your doubts and objections and gotten the President's verdict in my favour.'

He even sought the protection of the king, Charles I, with a flowery and flattering address:

> Most gracious King, The heart of creatures is the foundation of life, the Prince of all, the Sun of their Microcosm, on which all vegitivity depends, from whence all vigour and strength does flow. Likewise the King is the foundation of his Kingdoms, and the Sun of his Microcosm, the Heart of his Common-wealth, from whence all power and mercy proceeds.

But finally, after many attacks, Harvey's discovery was accepted across philosophical Europe before he died.

I present to you William Harvey, greatest Fellow of the Royal College of Physicians!

## Notes

1 *Essays on the Life and Work of Thomas Linacre* (1977), p. xv.
2 Charter as reprinted in Goodall, *The Royal College of Physicians of London* (1684), p. 7: 'Itaque partim bene institutarum Civitatum in Italia and aliis multis Nationibus exemplum imitati ...'; see also Charles Webster, 'Thomas Linacre and the foundation of the College of Physicians', in *Essays on the Life and Work of Thomas Linacre* (1977), pp. 198–222, see esp. p. 213: 'during his prolonged visit [to Italy] Linacre would undoubtedly have become familiar with the highly organized state of the medical profession in Italian cities', which included Padua.
3 In 1517 he published his version of Galen's *De sanitate tuenda* (On maintaining health), in 1519 his version of Galen's *Methodus medendi* (The method of healing), and later, in 1521 and 1522, of two of Galen's works on the temperaments of the human body, and finally in 1523 Galen's work on the natural faculties.

4  *An autobibliography by John Caius* (2018). p. 17.
5  Charleton, *The Immortality of the Human Soul,* (1657), pp. 34–43, see 34, 35. On this see Webster, 'The College of Physicians: "Solomon's House" (1967), pp. 393–412.
6  On which see Wharton (1996).
7  There was in fact one frequent occasion when a suspicion might have arisen in the mind of the doctor or the surgeon: that is, when preparing to bleed someone. Bleeding was practised on patients both to manage illness and to ward off illness too. Usually the blood was taken from a vein in the crook of the left arm, as it still is today if required. A ligature or tight bandage was tied around the upper arm, and this leads to the superficial veins in the lower arm swelling and becoming visible, so that the doctor or surgeon could easily lance a vein here and enable the blood to flow out. But if, as Galenic medical theory propounded, blood in the veins was flowing *outwards* from the centre of the body, then applying the ligature should in fact lead to the veins *above* the ligature to swell up, because their flow outward was now impeded, not those below it. In other words, in regular medical practice the veins swelled up on *the wrong side of the ligature*. However, this never seems to have been noticed by any doctor or physician before Harvey's day, and no suspicion about the two distinct blood flows in arteries and veins seems to have arisen from it. French, *The History of the Heart* (1979), p. 76. Harvey deals with the effect of ligatures in blood-letting, explaining what happens using the circulation of the blood in *Motion,* Chs 11 and 12.
8  On which see Minelli, *The Botanical Garden of Padua 1545–1995* (1995).

# Introduction

An English physician, William Harvey, sometime probably in 1618 in London, in the course of some research he was doing on animals, discovered that the blood circulates in the bodies of all animals. This is one of the most important discoveries ever made about the functioning of nature, and its consequences underpin all modern understandings of physiology and medicine both in humans and in animals more widely. It is, as it were, the foundation discovery on which the rest of physiology and medicine have been built ever since. Not only has it been the foundation for all subsequent thinking and investigation in these areas, but it also rendered redundant and erroneous the thousands of years of the thinking of philosophers and physicians, in the western tradition, about how the bodies of humans and animals function. It was a real turning point in the history of human thinking about the universe, as important subsequently in the life sciences as Newton's discovery of the laws of motion was for astronomy and physics.

I offer here a radically new interpretation of how and why the discovery of the circulation of the blood in animals was made. Questions about the nature and context of significant discoveries have long fascinated historians. What are the conditions under which individuals make novel and important discoveries about nature or come up with new and fruitful interpretations about nature? Much of the writing that we have about the history of science and medicine is concerned, directly or indirectly, with these issues. But there is still no consensus among historians of science and medicine on what the necessary and sufficient (or even the likely) conditions are which provide the context and impetus for new discoveries or interpretations. It is all too easy to assume, tacitly or explicitly, that it all comes down to genius – inspiration built on perspiration – on the part of the investigator. In other words, it is tempting to say that whilst the question is very interesting, it is nevertheless an unanswerable question, best left unasked. But, to my mind, to take that route is to renounce the possibility of ever understanding how new views about nature ever arise – which surely is one of the core questions in the whole of the history of science.

My own view on the making of discoveries centres on what I call 'projects of enquiry'. No-one can or does investigate nature naïvely and innocently.

DOI: 10.4324/9781003247616-2

We all necessarily come to the task already equipped with particular ways of looking, which we have acquired from our local cultural and social experiences.

My particular thesis in this book is that this discovery of the circulation of the blood in animals was made by William Harvey because as an anatomist he was following the model of anatomising first begun in the 4th century B.C. by the Greek philosopher Aristotle. That was his general 'project of enquiry'. Further, I claim that this particular discovery would not have been possible – for anyone - without this prior Aristotelian commitment. Thus I am claiming that being a committed Aristotelian anatomist was a necessary condition for the discovery of the circulation of the blood. To put it another way: this discovery, so absolutely fundamental to all *modern* medicine and physiology and their further development, was made by a man looking *backwards* for his inspiration, and a long way backwards at that – over 2,000 years! I also claim that the phenomena that Harvey saw and elicited experimentally in the animal body were phenomena which only made sense to an Aristotelian anatomist, and no anatomist committed to any other kind of programme of investigation would have seen or sought them. This contrasts with the customary view nowadays among historians of science and medicine that advances in the investigation and knowledge of nature in the 17th century were made by *rejecting* (rather than embracing) the doctrines and models of the great Ancient, Aristotle.

I believe that when he was engaged in anatomical investigation Harvey was pursuing what I call 'the Aristotle project'. In the present case I have therefore tried to establish that Aristotle had been an anatomist, and to specify precisely what kind of anatomist he was. No-one has attempted this before.

I have been fascinated by Harvey from the first day when as a student I was introduced to his 1628 Latin text in which he announced his discovery of the circulation of the blood in animals – simply the most important anatomical discovery ever made, and one with such vast physiological implications. Fifty years on I am still somewhat in awe at the scale and scope of this discovery, especially as it was the work of a single, isolated investigator. It was clear very early to me from what Harvey wrote and how he behaved that he had not been trying to discover the circulation of the blood, but was intent on some other question. It turned out, on further investigation on my part, that in fact no-one had ever tried to discover the circulation of the blood – and no-one had even thought that the blood might circulate. So over these years I have kept asking: 'what was Harvey's project of enquiry?' – what question (or questions) was Harvey pursuing, in the course of which he *quite unexpectedly* discovered the circulation of the blood?

Surprisingly other historians have not asked this question, they really haven't (see Chapter 10 in this book). For Harvey's discovery of the circulation of the blood has customarily been treated as a topic in the 'history of ideas', and that meant that it has been treated by historians as the final development of a sequence of ideas held or developed earlier by other people. Almost all the historical writing about Harvey has been written in this way, with historians tracking down what they considered to be the essential *elements* of the 'idea' of the circulation of the blood, and trying to determine who before Harvey had (supposedly) first come up with each of those observations or findings, and which elements Harvey added in order to make the 'idea' complete.

Because Harvey's discovery was treated in this way by historians, no question arose for them about precisely how or why Harvey had made it, because he was seen only as the person who gathered up and fitted together all the ideas of earlier people – the ideas which were (supposedly) necessary for Harvey to know in order for him to make the final dash to the discovery – as if it was some kind of race with an agreed goal. Some historians saw him as a genius in being able to do this, especially in his use of sharp observation and experiment, whilst others (such as William Hunter) saw him only as making the last logical step in an almost-solved puzzle, and therefore an investigator of no great consequence (see Chapter 10). But at all events the question of precisely what Harvey was doing never arose, since it was assumed that we already knew, viz. that he was trying finally – at last! – to put earlier people's ideas together properly in order to discover the circulation of the blood. The conventional historical studies on Harvey are thus full of Harvey's supposed 'predecessors', 'precursors' and 'forerunners' – an approach which assumes they were all trying to do something similar.

Another feature of Harvey's text that struck me early on was not only that he does not say that he was trying to discover the circulation of the blood, but neither does he locate himself in a tradition of people trying to discover the circulation of the blood. He does not even mention most of the characters who people the customary story of the discovery told by historians today.

In fact Harvey puts himself and his work into a quite different tradition: that of his own teacher at Padua, Fabricius ab Aquapendente, and, way back beyond him, that of Aristotle. And these two are the *only* people whose work in this area he treats as a model in his own anatomical work on animals. These were always treated by him as his two greatest guides. This is what he says to the reader at the conclusion of his Preface to his *Exercitations on the Generation of Animals* (1651):

> know, I tread but the steps of other men who have lighted me the way, and (so farre as is fit) I make use of their notions. But in chief, of all the *Ancients*, I follow *Aristotle*; and of the later Writers, *Hieronymus Fabricius ab Aquapendente*, Him as my *General*, and This as my *Guide*.

In my view the whole story of how and why William Harvey discovered the circulation of the blood is encapsulated in his phrase here: 'I follow Aristotle ... as my General, Fabricius ... as my Guide'.

It is well known that Harvey was an Aristotelian (like almost all his contemporaries), and that he worked within the Aristotelian world-view. It has also been long known that Fabricius was Harvey's teacher of anatomy when Harvey was a student in Padua in his youth, and that he too was an

Aristotelian, both as a philosopher and as an anatomist. But no historian has yet taken seriously this assertion by Harvey that he took Aristotle as his General and Fabricius as his Guide in anatomising.

One of Harvey's friends, John Twysden, in discussing Aristotle, recalled that Harvey had 'oft to myself said ... that he [viz. Aristotle] was the most rational and acute Philosopher that ever lived, that his writings were neer divine, that he never met with any thing in Philosophy, of which he met not some track in him ...'.[1]

'I follow Aristotle ... as my General'. A General (*dux* in Latin) is a military leader holding the highest rank, and whose role is to set military goals for his soldiers, to plan and to decide on strategy, and to lead his men successfully into battle and to make conquests or repel attacks. The men look up to and trust the General, and their role is to follow and obey him, and to fight in order to achieve the goals the General sets, trusting him to take them on to victory and glory. This is the sense - literally a military sense – in which we need to see Harvey's devotion to Aristotle: Aristotle has set the goals of anatomising and also laid out the strategy to achieve them, so he should be followed as a wise leader, taking us on to victory.

And when Harvey here says he is following Fabricius as his Guide (*praemonstrator* in Latin, literally 'someone who shows the way ahead'), he means that Fabricius is his Guide to achieving the goals that the General has set: he has shown the way to follow the General's plan successfully, so as to achieve the goals, the victories and conquests that the General had decided on. Harvey deliberately and willingly adopted the same General and the same Guide in his work on the motion of the heart and blood in animals, and in all other of his investigations of which we have trace.

So in my research I turned back to these two earlier investigators, searching for features of their work that Harvey might have been adopting and using to shape his own outlook and work in anatomising.

Fabricius was Professor of Anatomy at Padua University from 1565 to 1613. Harvey was a medical student in Padua and heard Fabricius' lectures towards the end of Fabricius' career. There is no evidence that they ever met much beyond this, and certainly none that they were 'great friends' (as has been suggested by one historian). But I found that no historian had done any substantive work on the anatomical investigations of Fabricius, even though he had published much of it in a series of booklets which were together intended to make up what he called *The Theatre of the Whole Animal Fabric*. So this was a gap I had to fill myself. And when I did so, I found that, in his turn, Fabricius said that he was modelling *his* own work on that of Aristotle – and only that of Aristotle!

But when I came to Aristotle and his 'animal books' (as I shall call them), I found that modern historians and philosophers – for Aristotle is still a very live figure in university philosophy faculties – had turned him into a complete Modern. He was credited with practising many disciplines which weren't actually to be invented for hundreds of years – biology, comparative anatomy, and so on. Obviously, going down that route would mean going round in circles. So this meant that I had to look at Aristotle and the investigative project recorded in his animal books anew, too, to see what it was that was later so attractive to Fabricius and Harvey that they could adopt it as a research programme for themselves.

I was not, of course, the first person to notice that Harvey claimed to be an Aristotelian, not at all. After all, Harvey goes on about his devotion to Aristotle at length in his book, *On the Generation of Animals*. But this is a book which historians have generally dismissed, and they have certainly not seen it as being in the same league as the book on the motion of the heart and blood in animals. The problem here was one of dating: the book the historians found valuable and modern even, was published in Latin 1628, whilst the book with the explicit praise of Aristotle only appeared in Latin in 1651. That is to say, it looked to some historians as though the second one was the product of medieval thinking, with Harvey's reasoning blinkered by his commitment to Aristotle, whilst the first one looked as though it was the product of modern thinking, rejecting Aristotle and embracing experiment. Had Harvey had a relapse and gone from being a Modern to being a slavish adherent of the Ancients? I argue here that all the anatomical research on which the two works were built was actually of a piece. If we look carefully, we will find that the 'modern' work, on the movement of the heart and blood in animals, was as much indebted to Aristotelian ways of thinking as was the later work, the book on the generation of animals. In both books Harvey is following Aristotle.

So all these questions on my part led me back first to Fabricius, whom I found claimed to be patterning his own work on that of Aristotle, and then back to Aristotle himself.

Given my starting-point – that I was now looking for something programmatic in Aristotle's animal books, programmatic enough for later people to model their own anatomical research on it if they chose – my reading of what Aristotle was doing in the animal books differs from that of almost every other modern scholar. The important thing here for me was to put aside our usual casual and unthinking ascriptions of modern disciplines to Aristotle, not just because they are anachronistic and misleading, though that should be sufficient reason, but because they close off the important questions just at the point where they should be being asked. So, absent from my account of Aristotle are Aristotle the supposed biologist, comparative anatomist, experimental physiologist, classifier of animals, and so on. Instead, I shall be trying to ask: what was Aristotle doing in the research recorded in his animal books, what was his 'project of enquiry'?

So I shall first look at the General and his plans and ambitions. I am going to outline the goals and strategy of Aristotle's investigations into animals, together with something of his views on how one can obtain reliable knowledge about things in the world. Both these areas were later fundamental to Harvey's work, as Harvey himself makes abundantly clear. It will become apparent that this produces a new picture of Aristotle himself and the work recorded in his animal books.

The historian Pierre Pellegrin has recently, and very wisely, pointed out that we often make false assumptions about past people – in this case Aristotle - and ourselves as all being engaged in pursuing 'eternal problems':

> It is irrelevant to think of [Aristotle's] biology as incomplete; rather it is radically foreign to us: produced in a world that is gone, it tried to answer questions that we no longer ask.[2]

Pellegrin also points out that most of the history of ideas has been predicated on this assumption that there are 'eternal problems', and that the historian's role is therefore to trace successive attempts by past people to solve these supposed 'eternal problems'. But if we follow Pellegrin in recognising that there is actually no 'identity of problems', then we can also follow his advice that in our historical research, before we join up two paths into one path – between say Servetus and Harvey – we need to make sure that they are the same path!

In order to understand Aristotle and his animal books our challenge here – and it is not easy – is that we have to come to him without liking or not liking, without preconception, without a desire to modernise him or to recognise ourselves and our modern disciplines in what he was doing. In Pellegrin's words, we need to explore Aristotle's animal books as a product of a man in 'a world that is gone' trying 'to answer questions that we no longer ask'. And later we shall have to look at what Fabricius and then Harvey made of these books, in another world that is lost to us, as they also tried 'to answer questions that we no longer ask'.

As should now be evident, my questions about Harvey, his discoveries and how he came to them, are very simple, direct and obvious: what was Harvey doing in the course of which he discovered the circulation of the blood, and why was he doing it? These questions serve to recapture the *identity* of past activities in the worlds and lives of past people. These questions apply as well to the history of ideas (which has been my own main field of interest) as they do to the history of politics, of social planning, of the writing of history in the past, or whatever. They are *open* questions, in that they do not presume any particular identity of activity on the part of our past investigators (e.g. in Harvey's case, that he was 'doing biology'), nor do they presume that our past people were trying to achieve or discover whatever it is that we now celebrate them for (e.g. again in Harvey's case, 'trying to discover the circulation of the blood').

But whilst they are clear and open questions, they are not usually particularly easy to answer, since in most cases we not only have to shed our preconceptions and presumptions about that past person's activity, but we also have to acquire the ears to hear just what it was they were saying to their contemporaries about what they were doing and why they were doing it. Nor can the two questions be raised and answered one at a time: as we shall discover in the present case, *what* Harvey was doing was of a piece with *why* he was doing it.

But on the other hand, in fact, these are usually quite easy questions to answer, because the past people we are looking at more often than not did specify the identity of what they were doing, and also announced why they were doing it. However, since these specifications and announcements were being made to their contemporaries rather than to us, we often find it hard to notice them as we look backwards in our role as historians. We shall find this to be true about Harvey too: he repeatedly specifies the enterprise he is engaged in – in the course of which he unexpectedly discovered the circulation of the blood – and says why he was doing so. And once we have ears to hear Harvey saying this, then everything else will fall into place.

## Notes

1 Twysden, *Medicina Veterum Vindicata* (1666), p. 47.
2 Pellegrin, *Aristotle's Classification of Animals* (1986), p. 2. Whilst Pellegrin here still unfortunately talks of Aristotle's 'biology', his discussion shows that he is well aware of the historiographical issues at stake here.

# 1 Aristotle's animal and the question of the soul

As can now be seen from my Prologue chapter 'Nine years and more', my story about how Harvey discovered the circulation of the blood begins by privileging Harvey's own claim, 'I follow Aristotle'. My argument then starts with Fabricius in 16th-century Padua trying to revive and to continue practising Aristotle's investigations into animals. Harvey, in turn, takes up this same practice of investigating animals, and takes things further than Fabricius had done. Finally I claim that, by following Aristotle both as practitioner and theorist, Harvey finds himself investigating an anatomical area that no-one – not even Aristotle, not even Fabricius – had investigated before, that is the heart and blood vessels of live animals (all animals, not just humans), but doing so without exploring the lungs or other respiratory organs (as all anatomists, intent on humans, had done before him). By complete immersion in this project over a number of years, by creating new experimental procedures, by dogged work and persistence, by refusing to be put off by the extraordinary findings he was making, Harvey discovered – not willingly, not with any sense of triumph – that the blood circulates continuously in animals, including man, while they are alive, and that when the heart stops pumping the blood then the animal dies.

'I follow Aristotle', Harvey wrote. But what Aristotle is this? And this Aristotle, what is his project of investigating animals? It is not an Aristotle we are familiar with today, nor are we familiar with this project of investigation by Aristotle, despite generations and generations of scholars exploring and discussing his work.

The standard explanation from Antiquity of what Aristotle was doing with respect to animals, and which is recorded in his animal books, comes from the elder Pliny's *Natural History (Naturalis historia)*, written in the mid-1st century A.D. Will this be of any help as we try to answer our question about the identity of Aristotle's project in investigating animals? Pliny is talking about lions when he writes:

> *Aristotle* ... a man whom I cannot name but with great honour and reverence, and whom in the history and report of these matters I mean for the most part to follow. And in very truth King *Alexander* the Great, of an ardent desire that he had to know the natures of all living

DOI: 10.4324/9781003247616-3

creatures (*naturas animalium*), gave this charge to *Aristotle*, a man singularly accomplished with all kinds of ... learning, to search into this matter, and to set down the same in writing: and to this effect commanded certain thousands of men, one or other, throughout all the tract as well of Asia as Greece, to give their attendance and obey him: to wit, all Hunters, Falconers, Fowlers, and Fishers that lived by those professions: Item, all Forresters, Park-keepers, and Wariners: all such as had the keeping of herds and flocks of cattle: of bee-hives, fish-pools, stews, and ponds: as also those that kept up fowl tame or wild, in mew: those that fed poultry in barton or coup: to the end that he should be ignorant of nothing in this behalf, but be advertised by them, according to his Commission, of all things in the world. By his conference with them he collected so much, as thereof he compiled those excellent books *de Animalibus, i.e.* of Living creatures, to the number almost of 50. Which being couched by me in a narrow room and brief summary, with addition also of some things which he never knew, I beseach the Readers to take in good worth: and for the discovery and knowledge of all Natures works, which that most noble and famous King that ever was desired so much to know, to make a short start abroad with me, and in a brief discourse by mine own pains and diligence digested, to see all.[1]

So Pliny here ascribes the whole project to an initiative of Alexander the Great '*of an ardent*' – if unspecific – *desire that he had to know the natures of animals* [so in the Latin, *all living creatures* in the English], who put Aristotle in charge of a grand investigative programme with thousands of men to work for him. Pliny gives no timing for this project so scholars have chosen, according to their lights, to place it at the beginning of Aristotle's career, or in the middle, or at the end, whichever best suits the story they want to tell about Aristotle's work.[2] Pliny here also indicates that his own project in the *Natural History* is of a similar kind and with similar goals. For, whatever source Pliny had for this story, he is clearly describing Aristotle as an early Pliny:

As touching myself (forasmuch as *Domitius Piso* says, That books ought to be treasuries and store houses indeed, and not bare and simple writings) I may be bold to say and aver, That in 36 books I have comprised 20000 things, all worthy of regard and consideration, which I have recollected out of 2000 volumes or thereabout, that I have diligently read ... and those written by 100 several elect and approved authors (Dedicatory Epistle to Vespasian, not paginated).

Where Aristotle supposedly had thousands of men to collect varied and unspecific information, Pliny has had 2000 volumes as his sources for his factual and imaginary stories. I think that Pliny's account of Aristotle's animal project can be trusted about as much as Pliny's own naive stories of elephants which speak Greek and climb trees upside down (*Natural History* book VIII, 1.3). Even if it is true, Pliny's account of Aristotle's project is so imprecise with respect to motivation, or to what information was to be sought, as to be of no use to us at all. So let us put this hoary old story aside and begin again. Our primary source has to be Aristotle himself, not some gullible fantasist such as Pliny writing several centuries later. It is, however, the case that Aristotle's animal books record and include the observations of very many people over a very long time, all brought together to serve as data for Aristotle's own questions.

So, what had Aristotle been doing when he was investigating animals, back in the 4th century B.C.? Why had Aristotle turned to them in the first place as an object of enquiry? For he knew of no predecessors in this enquiry, and we know of no systematic investigations into animals in the Greek tradition before Aristotle either. What was his enquiry about? What was he looking for? What questions was he asking? What kind of answers was he satisfied with? How did he demarcate his area of enquiry: precisely what did he take as his material for investigation? And quite how did he go about it?

Modern scholars of Aristotle's animal books have investigated virtually none of these questions. For they have worked with the assumption that they already know what Aristotle was doing in his investigation of animals: they take him to have been practising modern disciplines such as 'biology', 'comparative anatomy', 'taxonomy', 'embryology' and so forth (see Chapter 9 below). But we need to treat it not as a settled, but as an open question what Aristotle was up to: and to discover the answer to this we need to listen to what Aristotle himself said about what he was doing and why he was doing it. There can be no better witness than Aristotle himself to the identity and coherence of his own activity.

Today we think of Aristotle as a philosopher who could wear many hats: and we do not remark that the extension of his concept of philosophy and of the role of the philosopher is matched by no modern category of knowledge and by no modern scholar – and indeed is not even captured by all the faculties of a modern university and all its academics, taken together. In fact, the *modern* discipline of philosophy is itself a creation of the 18th and 19th centuries. The point, concerns, topics and identity of old philosophy were all transformed and reshaped then, and were greatly restricted compared to their former range and extent. The modern philosopher was now obliged to yield his expertise with respect to Nature to that new person, the scientist. The new philosopher now had to pursue his philosophy not in the world but in the ivory tower of academia: philosophy was no longer to be the pursuit which enabled one to live the life of the good man, but was restricted to a set of discrete topics for discussion and argument.

In order to appreciate Aristotle's view of the role of the philosopher and of the goals of philosophising, and of his view of the role of the study of animals, the first thing we have to do is to abandon the modern discipline boundaries and definitions. For it is only our commitment to these which leads us to characterise his animal books currently as 'biology' and so on, and therefore to contrast, oppose or even seek to reconcile them with his supposedly distinct and supposedly contrasting 'philosophical' works. *Our* discipline of philosophy has no role for the systematic study of animals. But it will become clear that *Aristotle's* philosophising was undertaken in order to understand the nature and functioning of the soul, and that this meant that an integral and crucial part of philosophy *had to be* animals. Only when we can see the animal books as an integral part of Aristotle's particular view of what philosophy is and how it should be lived and practised, will we have gone some way to understand them.

*Anatomising* is a means to an end, not an end in itself. The anatomist engages in anatomising in order to answer questions he or she has in their mind. The questioning mind guides the knife in the hand. The anatomist may wish to discover what kinds of different material the human, animal or plant body is made up of. Or he may wish to know how particular organs within the body work or carry out their functions, or even what these functions are. Or she may be inquiring about the similarities or differences between different animals, maybe asking why they exist, and perhaps even asking whence these similarities or differences derive. Or they may be seeking to find out how one complete animal or plant produces another like itself.

So: anatomy is a *practical and empirical* pursuit, aimed at acquiring *theoretical* knowledge. Without his knife in his hand, the anatomist is helpless: he cannot be an anatomist unless he cuts up and explores, as at least one part of his investigative procedure. But equally, without thinking and reflexion, the anatomist gains no more knowledge through all his cutting-up than the butcher has. This theoretical, contemplative, dimension of anatomising is what distinguishes the anatomist from the slaughterer, butcher, cook or huntsman, who seek knowledge of the insides of animals for practical purposes, such as preparing meat for table. While anatomical knowledge can certainly be put to practical use in other contexts, such as medicine and surgery (two of the disciplines in respect of which we usually think about anatomising), and this ultimate practical use may be the reason why any given individual engages in anatomising, yet in itself anatomical knowledge is not practical but wholly theoretical. This characterisation of anatomy applies equally to all the different anatomical projects that there have been over the centuries.

It is clear that Aristotle did actually perform anatomy, asking questions about the nature (and natures) of animals and their organs, questions which could only be answered by employing the manual practices of dissection and vivisection. The term 'anatomy' is ancient Greek in origin (rather than modern-style Greek, like 'biology'). There is extensive evidence in the animal books that Aristotle had personal experience of the dissection of a wide range of animals. Aristotle repeatedly refers his readers to his own (now lost) book, *Dissections* (or *Anatomy*). In the *Parts of Animals*, for instance, when speaking of the parts which pass the food, Aristotle writes:

The mouth, then, having done its duty by the food, passes it on to the stomach, and there must of necessity be another part to receive it in its turn from the stomach. This duty is undertaken by the blood-vessels, which begin at the bottom of the mesentery, and extend throughout the length of it right up to the stomach. These matters should be studied in the *Dissections* and my treatise on *Natural history* [i. e. the *History of animals*].

(650 a 28–32)

When discussing the blood vessels, he writes: 'For an exact description of the relative disposition of the blood-vessels, the treatises on *Anatomy* and the *Researches upon animals* should be consulted' (668 b 28–31). Similarly, when discussing the lobster, Aristotle writes, 'For an account of every one of the parts, of their position, and of the differences between them, including the differences between the male and the female, consult the Anatomical treatises and the *Inquiries upon animals*' (684 b 2–6). And everywhere in the *Parts of Animals* there is material presented which could only have been established by dissection; for instance: 'In all cases that we have examined the heart is boneless, except in horses and a certain kind of ox' (666 b 18–20). In other treatises linked to the animal books too there are direct references to dissection as in *On Respiration*, for example, and *On Sleep*. It is also clear that Aristotle had prepared diagrams for his *Dissections*. Speaking of the cuttlefish in *History of Animals*, for instance, Aristotle writes: 'The cuttlefish has two sacs and numerous eggs in them, like white hailstones. For details of the arrangement of these parts, the diagram in the *Dissections* should be consulted' (525 a 6–8).

Anatomy has two interdependent facets: manual (dissection) and mental (rational and contemplative). We have just seen that Aristotle engaged in the manual practice of dissecting animals. But why? What were the rational and contemplative grounds for him doing so? Usually this question would not arise in our scholarly work on him. And if the question were raised, it would seem like a question whose answer is self-evident: for we would assume that Aristotle obviously anatomised animals because in conducting the work recorded in the animal books he was being a biologist (or some other kind of modern investigator). In such a role or roles *of course* he would have anatomised animals. How else could one sensibly be a biologist and so on? But as that was not – *could not* have been – the case, the question can indeed be asked: what reason did Aristotle have for engaging in such systematic dissection of animals? – for it is not the sort of thing one engages in for no purpose.

In seeking to recover what Aristotle's object and project of investigation was with respect to anatomy, let us use Aristotle's own words as much as possible. And let us start our investigation of the nature of this project at the same point that Aristotle starts his own exposition of it – and by this I mean the beginning of Aristotle's *argument* about the study of animals, which may or may not be chronologically the earliest of his works in this area.[3]

This starting point is in the book *On the Soul*, known to scholars for centuries under its Latin title, *De Anima*.[4] We need to read it as the necessary preliminary to the animal books. Indeed it is more than a preliminary: it is the book whose theme and thesis called into existence the animal books themselves, and the work recorded in them.[5]

A work on the soul ought to seem to us to be a mighty odd place for anyone to begin the exposition of a project on anatomy. In the modern way of going about things, there is no relation between the 'soul' and the enterprise of seeking anatomical knowledge. Yet Plato, Aristotle's own teacher, talking in the *Timaeus* about the human body, had described it as divided into certain regions which correspond to the nature and needs of the soul, and talked of the body as 'the vehicle of the soul'. Indeed it can be claimed that the philosophical projects of each of that great triad of Athenian philosophers – Socrates, Plato and Aristotle – were soul-centred, that their respective philosophical systems are about the soul. Scholars have of course been right to point out the great differences between the outlooks of Socrates on the one hand, and his pupil Plato on the other; and between Plato on the one hand, and his pupil Aristotle on the other: there are indeed differences, and they are highly significant. Yet the things that Socrates, Plato and Aristotle disagree on, in successive generations, are the same things. That is to say, they agree on what the enterprise of philosophy is about: it is about the human soul gaining wisdom and understanding. They disagree on precisely what kind of thing the human soul is, and hence on what kind of wisdom and understanding it can properly acquire and how it acquires it.

In the modern world, however, we have divided up the topics of knowledge in a way which would have made no sense to Plato or Aristotle. In the first place we take the study of the 'soul' (insofar as it is thought to exist at all) to be the province of only one branch of knowledge: theology/religion. Similarly we have put just some of the concerns of the Greek philosophers into the subject-area that we now call 'philosophy': logic, metaphysics, epistemology, ethics, aesthetics. And others of their concerns we have placed in *our* subject-area, 'science': anatomy is one of these, which we count to be part of 'biology', itself regarded by us as a part of science. These subject-boundaries we work with are antithetical to the Greek philosophical enquiry. The soul and anatomy, which Aristotle studied as one, we have not only put into different departments of knowledge, but into ones to which we accord different statuses: theology/religion for us deals with 'revealed' truth, while science we take to deal with 'objective' or physical truth.

The defining of 'soul' as something which is the concern of theology/religion comes from Christianity being our paradigm of what a religion is. Yet 'soul' became a central concern of Christianity in the first place because of the influence on Judaism of Greek ways of thinking. Soul as the true man, or what-makes-a-man-what-he-is; the soul as something immortal; the soul as an entity separate and separable from the body; the soul as something implanted in the human body before birth; the soul as something which 'ascends' out of the body after death. All this is Greek and common to

Socrates, Plato and Aristotle. But we can also find all this about the soul in certain books of the Old Testament and many books of the New Testament because there was an extensive historical encounter of Judaism with 'Hellenism' – with, that is, the legacy of Socrates, Plato, Aristotle and others, and most especially of Plato. It was the career, safety, health, well-being, self-expression and fulfilment of this soul that the philosophy of Socrates, Plato and Aristotle was centrally about. The Christian soul was originally the soul of the Greek philosophers. But it was the soul's aspiration to, and its 'being at one with' wisdom (*sophia*) which defined and constituted the enterprise of Greek philo-sophy ('love of wisdom'). The Christians, however, gave this soul new goals, religious goals, exclusively super-natural goals: to seek eternal salvation, and to sustain a direct relation with God. Yet the old Greek philosophical goals for the soul can still be seen to a certain extent in the contemplative and ecstatic traditions of the Christian Church. And on the other hand, the fact that 'soul' is not a concern of our discipline of science or its sub-disciplines is a result of much later encounters with this same Greek philosophical tradition. For, when 'natural philosophy' was deliberately reconstituted as 'science' at the beginning of the 19th century, it was ensured that a central defining feature of this new discipline 'science' was that it most certainly did not and never will deal with either 'God' or soul.

When Aristotle, first as a pupil of Plato in the Academy and then as an independent teacher in the Lyceum, set out to engage in the enterprise of philosophy, he had to start (logically, if not chronologically) with the issue of what kind of thing the soul is, how it acquires knowledge and to what extent that knowledge is to be trusted. The *De Anima* 'On the Soul' is where Aristotle starts his own discussion of this topic. Or rather, it is where Aristotle starts his argument about the nature of the soul. For Aristotle, of course, knew the answer, in general outline, if not in all its ramifications and details, before he started the teaching or writing which is recorded in this book. Now, what Plato taught (and hence what Plato taught to Aristotle) was a view of the soul as a thing in three parts: an immortal soul located in the head; and a mortal soul, in two parts, located in the thorax and abdomen respectively. Plato was mapping the characteristics of this tripartite soul onto what he knew of the body of man. He was not doing anatomy. Now Aristotle, for reasons we shall explore soon, disagreed with Plato about the nature of the human soul: he disagreed with Plato's view of what kind of thing the soul is, what kind of thing it can 'know', how it comes to knowledge about things, the status of that knowledge, and the practical consequences for man of the wisdom that the practice of philosophy produces. Until Aristotle had specified his view of what the soul is and how it works, his attempt to reorient the philosophical enterprise could not get started. For what prompted Aristotle to teach and write was, as much as anything, his desire to correct his teacher: if, as a Plato, you have the wrong concept of the soul, you will philosophise incorrectly.

As we shall now see, Aristotle's particular view of the soul, of what it is and how it functions, was to lead him to the study of anatomy. Plato's view of the soul had not led him in this direction; nor had anyone else's led them thither either. Aristotle's attempt to refute his teacher is what started this particular western tradition of anatomising that we are concerned with. And the precise 'thing' in nature which Aristotle anatomises is given its identity by the starting point of Aristotle's enquiry: his view of the soul and of what the soul can know. For, in Aristotle's view – and here he is in agreement with Plato – the human soul can only 'know', can only 'be at one with' *universals*. Hence when he turns to anatomising, what Aristotle anatomises is not man, nor is it animals and nor is it different kinds of animal. It is, instead, a universal: it is 'The Animal'.

One of the things that Aristotle disagreed with Plato about was whether *this* world was of interest and consequence to the true philosopher. Aristotle agreed with Plato that the perfect, the eternal and the stable were the true objects of the philosopher's quest for knowledge. But Plato had claimed that such things were not perceivable in this imperfect and transitory world, only in the world of Forms; for Plato 'the world of perfect Forms contains all that is truly real'.[6] Aristotle, by contrast, claimed that the perfect, eternal and stable could indeed be seen within the imperfect and transitory things of this world.[7] As has been often remarked, this difference between them is nicely captured in the famous 'School of Athens' fresco in the Vatican painted by Raphael in 1509–12, which portrays Plato pointing upwards to the perfect, unchanging heavens and the permanent 'Forms', and Aristotle pointing downwards to the things of the changing sub-lunar world, each thereby indicating that *this* is what the true philosopher should be concerning himself with.

Aristotle talks about his desire to turn philosophising away from obsessive concern with the eternal divine bodies in the heavens, and to direct it also towards things in the constantly changing sub-lunar region, especially animals. In a celebrated passage in *Parts of Animals* Book 1 c. 5., he claims that here we can find constancy in the midst of change. This passage is sometimes quoted as being Aristotle's rationale for practising biology,[8] a claim we can now put aside. What we need to do is listen to it in the context of Aristotle's philosophy as a whole.

> Of the beings such as are composed by Nature – the ones we call ungenerated and imperishable for all eternity, and the others we say partake of coming-to-be and passing-away – it so happens that while the first ones [i.e. the heavenly bodies] are worthy and divine, there are very few views (*theoria*) of them available for us because, if we try to investigate these beings – even those which we long to know about - that which is clear to perception is very scanty. It is easier for us to acquire knowledge about the perishable plants and animals because we live among them, since one can acquire much knowledge about each kind that exists if one cares to take enough trouble. And each of them has its charm. For even if we grasp only a little bit about the heavenly bodies, yet by the excellence of the knowledge it is more pleasurable than all the things around us

(just as seeing a fleeting glimpse of our loved ones is more pleasurable than seeing many other things large and clearly), nevertheless with respect to the others [i.e. the earthly beings] since we know more and better about them they have superiority in knowledge [over what we know of heavenly things]. And furthermore, by being nearer to us and more similar in nature, they make some compensation for the philosophy about heavenly things.

As we have already treated the heavenly bodies, giving our opinion about them, it remains to speak about animal nature leaving out nothing as far as possible, neither nobler nor less noble. For even with animals which are not pleasant to look at, nevertheless the Nature at work in them holds extraordinary pleasures for those who are capable of recognising the causes (*tas aitias*) and are philosophers by nature. After all, it would be unreasonable and absurd if looking at pictures of them we rejoice that we see [both the representation of them and] at the same time an art-at-work (such as the art of painting or the art of sculpture), and yet we did not love more the contemplation (*theoria*) of the things themselves composed by Nature since we are capable of discerning the causes.

Therefore one must not be childishly disgusted by the study of less worthy animals, because in all the things in Nature there is something marvellous. And just as it is said that Heraclitus said to some visitors who wanted to meet him and stopped when they entered and saw him warming himself by the stove - he ordered them to come in and not be afraid, 'because even here there are gods' – so one must go about the investigation of every animal without expressing distaste, since in all of them there is something of Nature and of The Beautiful. For the non-random, the for-something's-sake is in the works of Nature most of all; and the thing because-of-which it is composed, or has come-into-being – its purposes (*telos*) – are part of The Beautiful. And if anyone thinks that the investigation (*theoria*) of the other animals is unworthy, then he must think the same way about himself too, for it is not possible without much disgust to see the things of which the human kind is composed, such as blood, flesh, bones, veins, and other such parts. Equally, one must recognise that he who discusses any one of those parts or equipment is not speaking about the material (*hyle*) itself and not for its own sake, but for the sake of the entire form (*morphe*) – just as one discusses a house but not the bricks, mortar, timber. In the same way one must recognise that discourse (*logos*) about Nature is about the composite-thing and the entire being as a whole, but not about those things which never occur separately from those beings. (644 b 22 – 645 a 37; 645 b 15–20)

… First of all, our business must be to describe the attributes found in each group; I mean those "essential" attributes which belong to all the animals, and after that to endeavour to describe the causes of them. (645 b 1–4)

... Now, as each of the parts of the body, like every other instrument, is for the sake of some purpose, viz. some action (*praxis*), it is evident that the body as a whole must also exist for the sake of some complex action. Just as the saw is there for sawing and not sawing for the sake of the saw, because sawing is the using of the instrument, so in the same way the body exists in some way for the sake of the soul, and the parts for the sake of those functions for which each of them has been formed. (645 b 15–20).[9]

# II

*On the Soul* opens with a delightfully disingenuous passage: in undertaking philosophy (the pursuit of wisdom), Aristotle is saying, an investigation into the nature and properties of the soul is of the greatest importance. It will be of especial use to the understanding of Nature since 'the soul is in a sense the principle of living things (*arche ton zoon*)'.

Aristotle's aim here, he says, is to discover (1) the nature and essence of the soul, and (2) its attributes/properties. And the first problem – indeed what may turn out to be the core of the problem – is to distinguish between (a) such properties as are characteristic of the soul by itself, and (b) such properties as cannot be separated from its presence in living creatures.

After this opening statement Aristotle goes on to raise a number of questions which (he implies) present themselves to anyone taking the soul as a topic of investigation. As with any good expositor, these questions are formulated in Aristotle's own terms: that is, he sets them up here as questions because they establish the terms of what he considers to be the proper answers.

But before we look at these questions we need to look at another issue which Aristotle raises here, and which is of great moment in our attempt to recover the nature of his investigation into animals:

> The knowledge of essences (the essential nature) of things is the only worthwhile knowledge; the essence of a given thing can be expressed in a formal definition. In some cases, such as mathematics, we can work from the definition to a demonstration of the attributes. In other cases we need to start by working from the attributes to the definition; only then can we demonstrate the attributes from the definition.

In his investigation of the soul (as in his investigation of everything else as well), Aristotle wants to arrive at the strongest way of showing that one knows something: this strongest way is being able to provide a demonstration. In a demonstration one starts from the definition of the essential nature of the thing in question, and then demonstrates (that is, shows deductively) that the attributes of that thing follow from that definition. They follow logically: and what follows logically of course follows necessarily – as its root in the word *logos* (word/reason) shows. Thus, if the definition of a given thing is x, then it can be shown to follow that it – all instances of it – have the attributes y,z,a,b. In showing that all the attributes follow logically from (and are contained logically in) the definition, one is also showing the causes of the attributes: showing why the things logically (necessarily) have the attributes they have. In the case of a mathematical definition, one can define a circle as being x; one can then show that it logically follows that it has the properties y,z,a,b. And one can assert with assurance that all circles will always have all these properties – because they follow from, are contained within, the definition. The definition is true of 'the circle': hence it is true of all circles.

But for Aristotle mathematics is somewhat of a special case: it is the only area of knowledge where one can start from definitions. When it comes to things in the world which really exist, then one cannot hope to start with the definition, and then show how the properties or attributes follow from that definition – because one simply does not yet know what the attributes or properties of those things are. So, in enquiries which concern things which actually exist in the world, one has to start by exploring the attributes the things actually have. Until one has done this, one stands no chance at all of arriving at an authentic definition. And we will have to look at every case we can, because we intend our definition ultimately to deal with, to be true of, every case (410 b 20): we want our definition to be such that we can demonstrate that all the important attributes that actually occur do indeed follow logically (necessarily) from it. Aristotle is not suggesting some crude form of induction here: he is not suggesting that we can get an exhaustive series of attributes which will, when distilled, turn out to be the definition we are seeking (a sort of lowest common denominator). No: what he is saying (as his language in the passage above reveals) is that only when we have looked as thoroughly as possible at all the instances we can find, will we have a sufficient clue or hint such that we will be able to create an adequate definition. But going from the attributes of the particular instances to the definition will still involve a mental leap. So: long before one can show the causes of a given thing demonstratively, one will have to become acquainted as thoroughly as possible with the properties or attributes of the things themselves.

It should be clear therefore what Aristotle's procedure is going to be. He wants to reach a proper definition of the soul, which will be a demonstration of its essence (of what it is); he will then be able to demonstrate what attributes the soul has, in all its particular instances, from that definition. This demonstration will itself be an account of the causes, the logical causes, why the soul has those attributes (why it is as it is). But he is going to start his investigation not at the definition, but at the attributes that the soul actually has in its manifestations. And that brings him back to his original question of establishing which attributes of the soul are characteristic of the soul itself, and which ones depend on, and cannot be separated from, its presence in living creatures.

Aristotle's answer to this question is built into his formulation of it: if any function or affection of the soul is peculiar to it, it can be separated from the body; but if there is nothing peculiar to the soul, it cannot be separated. Yet, Aristotle claims, in most cases it seems that none of the affections, whether active or passive, can, though possibly thinking is an exception. Thus (Aristotle concludes) the soul does not have any attributes of its own: there are no attributes of the soul except when the soul is present in a living body. We have already learnt that in our investigation of the soul we have to work from its properties or attributes, towards its ultimate definition. It is clear from the above finding that the soul has attributes only when it is, quite literally, 'embodied' in living creatures. Hence any definition we might give of any of the attributes of the soul must take this fact into account: and any attribute of the soul must be defined as a movement of a part of the body, or of a part or faculty of a body, in a particular state, aroused by a particular cause, with a particular end in view (403 a 27).

This is why the enquiry into the soul is the task of the philosopher of nature: for it is the task precisely of such a philosopher to deal with 'all the functions and affections of a given body, i.e. of matter in a given state' (403 b 12). The philosopher of nature deals with these two things at once: with matter, and with the state (or 'form') that it is in. The way that Aristotle has defined the issue, and the methodological considerations he has claimed are required in pursuing it, have turned the investigation of the soul into a matter of the philosophy of nature. No-one, Aristotle is claiming, can investigate the soul unless they start their enquiry from its actual manifestations in living creatures. The nature of the soul is the object of Aristotle's enquiry here: and it is this enquiry which has led him to look at living creatures. Aristotle looks at animals because he is interested in the soul. That is his motive; that is why animals have attracted his attention as a philosopher. And throughout all his surviving works which deal with animals, this motive remains the same for Aristotle. He does not look at animals for their own sake, for the sheer interest of what he might find. He looks at them solely with respect to what light they shed on the essence and attributes of the soul. He looks at animals because they are the soul-in-action.

Already (and we are only a couple of short passages into *De Anima*) we can see why Aristotle looked at and investigated animals at all, and we can also see what he looked at them as: he looked at them as instances of the soul-in-action. Surprisingly this way of seeing animals became enshrined in Latin, and hence in many modern languages which are based on Latin, such as English. The Latin for 'soul' is *anima*: an 'animal' is a 'thing which has soul'. Aristotle is interested in animals only insofar as they each and all embody the soul: the only way to study the soul is to look at it in action, in animals.

More strictly still, Aristotle is not interested in animals, but (if I may put it this way) in 'The Animal': the soul in its manifestations. For Aristotle 'The Animal' was a perfectly coherent object of study, which was a fully integral part of his philosophical endeavour. 'The Animal' as an object of study is neither a generalisation nor an abstraction in the modern sense of the term; it is, instead, a 'universal'. 'The Animal' comprehends man and brutes on completely equal terms: man is not a special object of study for Aristotle. Man is just another animal, just another instance of the manifestations of the soul.

Which brings us to the next section of the *De Anima*. In this section Aristotle proclaims that it is necessary, before he goes any further, to look at and review the opinions of his predecessors on this matter. We can be sure, however, that he will be reviewing their opinions in the light, and in the terms, of what he himself wants to prove a little later. This technique of presenting a seemingly innocuous 'review' of his predecessors' opinions is something Aristotle often recommends and practices at the beginning of his treatises. It is a practice in which Harvey was much later to follow him. These 'reviews' have been a major source for historians wanting to reconstruct the views of philosophers before Aristotle: but it has been brilliantly shown by Harold Cherniss that not one of these supposed 'views' can be read at face value.[10] For Aristotle always quotes other people as a means to the end of making his own case. And who can blame him? On this occasion Aristotle cites his predecessors on the question that he himself has raised (and in a way in which they most probably did not raise it!): the nature of the soul, and its attributes. Most revealingly, however, Aristotle says

> But our enquiry [into our predecessors' views] must begin by laying down in advance those things which seem most certainly to belong to the soul by nature. There are two qualities in which that which has a soul seems to differ radically from that which has not; these are movement and sensation.
>
> (403 b 25)

So it is on these issues that Aristotle is going to review his predecessors: how do they account for movement and sensation, the two basic attributes that Aristotle believes the soul obviously has? He plucks their 'answers' to these questions – *his* questions – from their discussions, which may well not have been treating these questions.

To cut a long story short, through his way of setting up the question, Aristotle can claim that 'there are three ways of defining the soul which have come down to us': (1) as the principal cause of movement; (2) as composed of the finest particles; (3) as composed of (an infinity of) elements.

Not surprisingly, Aristotle finds that all of these views have enormous limitations and suffer from being ill-thought-out and self-contradictory. Having discussed them at length, and strictly on his own ground, Aristotle is thus able to clear the decks and say 'let us start afresh, and try to determine what the soul is, and what definition of it will be most comprehensive'. We have reached Book II.

## III

Of the questions that Aristotle raised at the beginning of *De Anima*, and which he implied must face anyone taking the soul as a topic of investigation, the first was: to which of the genera the soul belongs, and what is it? Is it a substance, or is it a quality or a quantity, or something else? The second was:

does it actually exist or only potentially? At the beginning of Book II Aristotle raises these questions anew, and sets about answering them. He has by now, we must remember, dismissed the 'answers' that his predecessors had (or had not) given to them; the field is his own, and now the terms are exclusively his own. We must listen to Aristotle's answers to these questions since his answers lie at the base of everything he later does in looking at animals. The terminology is awkward for us, but must be faced, for these are the terms in which Aristotle (and later Harvey) expressed themselves, and which parallel the concepts with which they were thinking.

Let us look at the first question like this, Aristotle suggests. We can see that there are three aspects to substance:

(i)   there is 'matter';
(ii)  there is 'form' (the shape or structure);
(iii) there is matter-and-form conjointly.

Now, in Aristotle's view 'matter' never exists as such: it always has some form, and hence every physical thing that exists does and must consist of matter-and-form conjointly. However, we can have a concept of 'matter' without any form, and equally we can indeed have a concept of 'form' without any matter; and although we must appreciate that 'matter' and 'form' never exist separately from each other in the world (in the things which actually exist), as concepts they have a useful role in our reasoning about the things which do actually exist. Thus we can think of 'matter' as potentially being some 'real thing' or other; we can think of 'form' as what makes 'matter' an actual particular 'real thing'. 'Form', somehow added to 'matter', is what makes the things themselves what they are. Obviously it has to be 'form' added to the particular kind of 'matter' appropriate for that particular 'real thing', but it is the form which makes that thing what it is, and not the matter. If we have followed Aristotle's argument so far, we will realise that for living creatures the soul is what makes them what they are, viz. living creatures. Hence the soul is a substance, in the sense of being the 'form' of living things. Living things are matter-and-form conjointly; the 'form' in this conjoint entity is the soul.

We can join Aristotle in giving a preliminary definition of the soul as 'the first actuality of a natural body potentially possessing life' (412 b 1). Soul is what gives a seed – a natural body potentially having life – actual life, in the first instance. When soul is present, then a potentially living body is a living body: the potential has become actual. If the soul is absent (as when an animal dies), then the body is now an 'animal' only in name, for it has lost what made it a living body (an anima-l), its soul, its 'form'. This also helps us distinguish the sense in which Aristotle uses 'form'. When he considers the soul as the 'form' of an animal, he is talking about what-makes-it-what-it-is. As we can see in the case of a dead animal, particular structures or shapes of the matter, though essential to the functioning of an animal as an animal, are not in themselves the same as the 'form': they are necessary for the 'form' to be present and active, but they can be present – as mere structures or shapes – without the presence of the 'form' proper, the soul. The 'form' proper, thus, is defined by what it does, by the animal being animated, not by the mere structures in which the matter of any natural body is arranged. The structures (the arrangements of the matter) are the potential: the animal is an (live) animal only when these structures are actualised (made functioning) by the presence of the soul, the 'form'.

'The soul may be defined as the first actuality of a natural body potentially possessing life; and such will be any body which possesses organs'. Suddenly Aristotle seems to have gone beyond what his argument strictly leads to. What is this about organs? 'Organ' simply means 'instrument'. If we recall again that Aristotle's object of enquiry here is the soul, and we recall also that he has argued that none of the functions or affections of the soul is separate from the living body, we will begin to see why he adopts this way of conceptualising the body as divided into, or made up of, so many organs. For he takes it as given that the soul has a number of functions or affections: he has already mentioned anger, courage, desire and sensation generally, and he will later mention many more. These are all to him distinguishable functions that the soul has. It follows for Aristotle that the (animal) body must have certain bits which do something in order for each of these functions or affections to find expression (though not necessarily on a one-to-one basis). That is why he thinks of the body as consisting of so many organs – and they are instruments of the soul.

Armed with this assumption that 'organs' must be there in the animal body (simply because it is an animal), Aristotle was able to discover them there. Aristotle identified them on the basis of what they did, of their service to the soul – the soul which made that body an animal. Already, then, we know it is implicit in Aristotle's endeavour that the animal body could be 'anatomised' or cut up, and that certain ways of anatomising it would reveal 'parts', 'parts' which had 'demarcation by nature' according to the attributes of the soul. We can also anticipate that Aristotle will use vivisection in his enquiry, as well as dissection; for his concern is with the body functioning (the soul-in-action), and it is in vivisection that its functions are most clearly seen.

We come now to Aristotle's second main question: does the soul have parts or not? With this goes a subsidiary question: is every soul of the same kind, or are there different types of soul – one for each kind of animal, perhaps? Most people hitherto have begged this question, Aristotle claims (402 b), since they have only been concerned with the soul in man. In introducing this pair of questions Aristotle makes a point about the kind of definition he is after. A proper definition will not simply specify the fact: to define 'squaring the rectangle' as 'the construction of an equilateral rectangle equal to an oblong rectangle' merely states what a squared rectangle is. But to define it as 'the finding or discovering of a mean proportional' is a good definition, because it explains the cause why a squared rectangle is what it is – and it also explains how it comes to be a squared rectangle. Aristotle discusses this in order to make the point again that we have to start from the data of sense in seeking to reach an adequate definition. We have to ask how the soul operates, and understand why it does so, and what it is, before our definition will be adequate to embrace the what, the how and the why.

Does the soul have parts or not? And is there more than one kind of soul? The distinctive feature of the things which have soul is of course that they are living; and living involves (Aristotle says) the presence of mind, sensation, movement (or rest) in space, and the movement involved in nutrition, decay and growth. In that they have in themselves a capacity to feed, to decay and to grow, *plants* are alive. These capacities Aristotle describes as the 'nutritive': 'we call "nutritive" that part of the soul of which even plants partake'. This 'nutritive' aspect of the soul, what we might call its 'nutrivity' was later often called the 'vegetative' soul, or vegetivity; we shall see that William Harvey prefers this term. All animals have this set of capacities *plus* sensation (the sense of touch); some animals have both of these *plus* movement; and at least one animal (man) has all of these *plus* thought. The soul is the *origin* of – what is responsible for, the source of – all these faculties: nutrition, sensation, appetite, movement and thought, are all *aspects* of the soul. Of these, thought alone may perhaps be separable. But the other faculties cannot be thought of as 'parts' of a single soul (or as separate souls) for the evidence of sense is against it. For one can successfully take cuttings from a plant; and one can cut certain insects into two, and both parts will continue to live. The evidence appears to show that somehow the soul is a 'unity' which can be potentially divided into many little 'unities'; it is not a series of 'parts' joined and held together. The soul does not have parts, therefore.

This way of approaching the question of whether the soul has parts or not has brought Aristotle to the position where he can claim that life consists of various 'faculties' (*dynameis*), and the possession of even one of these indicates the presence and workings of soul; he has also claimed that they are simply aspects of the one soul. But what he has established in addition is that there is a hierarchy of these faculties: 'of the faculties of the soul which we have mentioned, some living things, as we have said, have all, others only some, and others again only one' (414 a 29–30). Given this variety in the extent to which the soul is present in different living things, it is clear that one definition of soul which covers them all is going to be too general: it 'would fit them all, but would be descriptive of no particular' one of them (414 b 25).

Following Aristotle, we therefore need to reformulate our question. Given that the soul has several aspects, or 'faculties'; and given that living things share to a greater or lesser extent in these faculties; given too that a definition of soul which covers all of these would be too general to be worth seeking, then we need, instead, to seek definitions of each of these faculties. Then these definitions, taken together, will be the full definition of the soul: they will cover all aspects of the soul as it actually has incidence in living things. We have noticed that these faculties are in a hierarchy, with plants at the lowest end possessing just the nutritive faculty, and man at the highest end possessing all of the faculties. This raises a further question which we must ask: why are living things (the different instances of the soul-in-action) thus arranged in such a series (415 a 2)?

Now that we have realised that the soul-in-action is a number of faculties arranged in a hierarchy, the tasks ahead of us in our pursuit still of a full definition of the soul are these:

1.   to explore each of these faculties (nutrition, sensation, appetite, motion, thought, and any others).
     To do this

     a)   in general; and
     b)   in particular, by exploring the soul (the particular set of faculties) of individual types of living creatures, 'for instance of the plant, the man, and the beast'.

2.   to investigate why these faculties are arranged in a series.

As Aristotle writes: 'It is clear that the account of each of these faculties is the most relevant account that can be given of the soul' (415 a 13–4). In all stages of this exploration we will be starting from the actions of the soul: from the faculties of the soul as actually exhibited in (particular) animals.

The third and last of the major questions is now raised: what should be our method or procedure of enquiry – where shall we start? Aristotle points out that before we can understand what each of the faculties are (the thinking, sensitive, nutritive faculties, for instance), we first have to explain what the functions 'thinking', 'perceiving' and so on are. To do this we need first to have understood what the 'objects corresponding to' these functions are, that is, what they think, perceive, and so on. And even before this, we need to determine 'the facts about those objects, e.g. about food, or the object of perception or (that of) thought' (415 a 22). At last we know where to start in this investigation of the soul: we must first start with things such as 'food', 'colour' and 'sound'!

## IV

Aristotle has now established, to his own satisfaction, that there are five major faculties of the soul: (1) the nutritive/vegetative; (2) the sensitive; (3) the appetitive; (4) the motive; and (5) the thinking. They are in a hierarchy, rising from the nutritive to the thinking. In the rest of *De Anima* he deals with these faculties in a general way and, following his own considerations on the appropriate method of procedure, he starts in each case with things such as food, colour and sound.

The first faculty that he will deal with will be the lowest, the one which all living things share, the nutritive/vegetative; it is through the presence in them of this aspect of the soul that all living things have life. This aspect of the soul has two major functions: to assimilate food, and to generate other living things of the same kind. After all, as Aristotle says, it is 'for the sake of' generating another living thing of the same kind that every living creature performs all its functions.

In mentioning 'for the sake of', Aristotle is now talking about cause; and thus he diverts for a moment to discuss all the senses in which this term can be taken. Hitherto Aristotle has been talking about living creatures because his quest is for the soul, and they are (for him) the soul-in-action. Now he turns the issue round, and asserts that the soul is the cause of living creatures. It is the cause in all three possible senses of the term:

1. it is the cause in the sense of what-makes-them-what-they-are, that is, living. The presence of the soul is the cause why any potentially living creature is an actually living creature. This sense of cause, the 'what-makes-a-thing-what-it-is' we have already met, and it is often translated as the 'essence' or 'essential nature' of a thing. In the case of living creatures, their 'essence' or 'essential nature' is their possession of soul.
2. the soul is also the cause in the sense of being the first principle of living creatures: it is the (internal) source or origin of motion in the living body. And of course, motion is characteristic of living creatures for Aristotle: (a) change of quantity (growth and decay); (b) change of quality (as in sensation); (c) change of place (locomotion).

The third, and most important, sense in which the soul is the cause of living creatures, is cause in the sense of that-for-the-sake-of-which. All living creatures exist for the sake of being instruments of the soul. This is a very, very important concept for Aristotle: the 'goal' (*telos*), the 'end', the 'for-the-sake-of-which' something exists. The concept, used in this particular way by Aristotle, was to be vastly influential. It is evident for Aristotle that the soul is prior to the bodies of living creatures, which are simply its instruments. He constantly draws a parallel with mind: humans act with some purpose in view, and Nature is just the same. Recognising this is the clue to understanding why Nature is as it is. For instance, if a man wants to make a table, he has this in view as his 'goal'. Everything he subsequently does in choosing and shaping the appropriate materials is for-the-sake-of-this-goal. The goal he has in mind is what controls and directs his actions. If we were to inspect the man's actions, then we can account for their coherence (and they do have a coherence) only by recognising that they are determined by the goal he has in mind. Now, the table comes into physical existence only at the end of the shaping process; yet as the intended goal of that process, it existed prior to that process, it is the cause of that process and it controlled that process. Thus the table as the 'goal' of that process is the 'for-the-sake-of-which' that process existed and was the process that it was. The same is true for the soul with respect to the body: the soul is the 'goal', the 'for-the-sake-of-which' the living body came into existence and exists. A living creature does not exist 'for its own sake': it exists 'for the sake of being' the instrument of the soul. The 'goal' of serving the soul thus controls and determines how any living creature must be, what parts it must have.

It is thus the requirements of the soul which specify and control what forms or forms living creatures can have. When he is looking at living creatures Aristotle is confident that their characteristics are determined (within certain limits, which will be discussed later) by the needs of the soul. And it is because the soul is 'that for the sake of which' living creatures exist, that Aristotle's project of finding out about the soul by looking at animals is both possible and logically sound. For animals are not just a convenient means of looking at the soul-in-action: animals exist simply to be instruments of the soul! They have, for Aristotle, no interest except as being instruments of the soul. We should look at animals because they are the instruments of the soul; and everything we find out in this enquiry will tell us more about the soul and its operations – because the soul is here the object of the enquiry. This is beginning to sound very like Plato's claim that the (human) body is simply the 'vehicle of the soul', and his insistence likewise that the body exists 'for the sake of' the soul. Aristotle is indeed dealing with this issue because he has inherited this way of thinking from Plato his teacher. But Aristotle is turning the issue upside down: he is saying that instead of assuming what kind of thing the soul is, what its requirements are and how the body must therefore function to fulfil these requirements, Aristotle is claiming that we need to investigate and hence *discover* the nature of the soul from the soul-in-action. This central concept, the 'for the sake of which', will be the grand clue to enable us to understand why animals are the way they are, why they have the parts they have and why those parts function the way they do. Investigating animals is the necessary means to the end of understanding 'The Animal' – which is 'The Soul in Action'.

Let us now turn, with Aristotle, to looking in a general way at the nutritive/ vegetative faculty ('the vegetative soul' as it was later referred to), the faculty common to plants and animals. This faculty involves a change of quantity of the living thing: preservation, growth and the generation of another thing like itself. It is achieved by a change of quality in which food (which has one nature) comes to be part of the living thing (which has a different nature); or, in Aristotle's terminology, a change from food to 'the thing fed'. This change cannot be due to matter (such as fire) as some people have suggested. It must be due to the operation of soul. It is soul, the agent by which the food becomes the thing fed, which preserves the living thing in its current state, and which thus enables it to generate another living thing like itself. It is the soul which also governs the growth of living things, such that they attain their final form through all the necessary stages, and reach and maintain their optimum, their final, form. Aristotle has here distinguished more closely the operations which the vegetative faculty embraces. It is implicit in what he has said that it will later be necessary to identify the appropriate organs through which these operations are carried out – the organs, that is, which serve the vegetative soul. At this point, however, he merely suggests that this series of changes probably involve heat, since 'that which produces digestion is heat; therefore everything which has a soul has heat' (416 b 30).

# V

Aristotle now turns to a general discussion of sensation, the second aspect of soul, the one which all animals possess, but no plants. Sensation, for Aristotle, is a change of quality, consisting in being moved and being acted upon. It is clear from this that Aristotle sees sensation as a passive process: the sense-organs are acted upon by something outside themselves, at which point their potential for sensing is made actual. It is thus essential, of course, that whatever it is that acts upon a given sense-organ actually exists: and here lies Aristotle's fundamental trust in the reliability of the information that the sense-organs give us. Now, the sense-organs are of one nature, while whatever it is that acts upon them is of another nature. As with the operations of the vegetative soul, therefore, here with the sensitive soul we are concerned with a change of nature, with how a thing of one kind (a movement of the external media) becomes another thing (sensation, i.e. the active, actualised, sense-organ). Aristotle is at pains to argue that the change of nature that occurs in a sense-organ is simply a change from its potential for acting to its actual acting: a 'realisation of its nature' (417 b 19); during the process of being acted upon the sense-organ is unlike, but at the end of the process it has become like that object (that it senses) and shares its quality (418 a 5).

There are five senses: sight, sound, smell, taste and touch (which last is, for Aristotle, located in a sense-organ lying just below the skin). To discover what is going on in sensation let us take the same two senses that Aristotle deals with here, sight and sound. A particular 'real thing' exists (for sensation is impossible without there being a real thing to be sensed). Its colour sets up a vibration in the air; or (in the case of sound) its vibration sets up a corresponding vibration in air or water. The medium, air or water, is continuous from the object to the sense-organ (the eye or the ear). At this moment the eye or ear is potentially sensing. During the process of 'seeing' or 'hearing', this vibration is transmitted to the organ itself, which by its nature is able to accept this vibration. In accepting it, the organ is made active: its capacity to act, its potential for acting, becomes actualised, and sensation takes place. It is in this way that 'during the process of being acted upon it [the sense-organ] is unlike, but at the end of the process it has become like that object [that it senses] and shares its quality'.

So it is not the 'thing itself' which is sensed by the sense-organ. It is a *quality* of that thing, such as its colour, which is made actual in the appropriate sense-organ. For this to be possible, it is necessary that the sense-organ be potentially all the qualities it can sense: the eye, for instance, is potentially all colours, and the ear all sounds. For discrimination to be possible (between white and black for instance) it is also necessary that the sense-organ be a 'mean' between the extremes of quality which it can sense. Everything that is sensed is sensed as being relative to this mean. Each sense-organ, therefore, is that in which this potentiality lies, and which has a 'mean': the presence of this potentiality and of this 'mean' together, is the sensitive soul. The sensitive soul is a 'first principle': it is the cause (in the sense of origin or source) of the operation of the sensitive faculty.

But what a given sense-organ senses is not the form-and-matter conjointly of a given 'thing itself' in the world. Each sense-organ senses a particular quality abstracted from the 'thing itself' being perceived. Each sense-organ receives information about one or other quality of the 'thing itself', and there are five different qualities that can be received about a given object by the senses: colour, sound, smell, taste and texture (including hot/cold). Thus each sense-organ is capable of receiving one aspect of the 'form' of the thing being perceived. Thus the sensitive soul is a 'first principle' capable of receiving the form without the matter of the 'thing itself' which is being perceived.

If we were to be using all our senses simultaneously in the perception of a given 'thing itself', we would be in receipt of the 'forms' of its colour, its sound, its taste, its smell and its texture. But that is all. The senses by themselves cannot perceive anything about the movement, shape, magnitude or number of the things sensed. That is, they cannot identify the 'thing itself' in its wholeness, as being what it is. To perceive the movement, shape, magnitude and number – and hence the identity and existence – of a particular 'thing itself', it is necessary that the data from each of the senses be brought together and assembled. This must take place in a 'common sense', in a united faculty.

What is this 'common faculty', this 'common sense'? Other philosophers (Aristotle claims) have implied that it is identical with the faculty of thought. But this cannot be, because all animals can perceive, while only some of them can think. The faculty in question is concerned with 'forming an opinion exactly corresponding to a direct perception' (428 b 3), and Aristotle identifies this faculty with the Imagination. The sense-organs are actualised by the presence of a 'thing itself'; the sense-organs transmit their motion, and thus 'Imagination must be a movement produced by sensation actively operating' (429 a 4). It is in the Imagination that the different aspects of the 'form' of a 'thing itself' come together from each of the sense-organs, into an image or picture. Just as the senses can sense (become actualised) only in the presence of a 'real thing' so, when it is working under the direct stimulus of the senses, the Imagination must be presenting an image of a 'real thing' in the world. We can trust the evidence of the senses, and we can also trust the evidence of the Imagination when it is actualised by the senses. Man of course (and possibly other animals) can also recall images at will, and these may or may not correspond to some 'real thing' or other in the world: in Imagination we are like spectators looking at something dreadful or encouraging in a picture (427 b 45), Aristotle writes. But the faculty of Imagination only involves looking at the picture. The next stage – judging whether or not it is a true picture, having opinions about the image – is the action of the next faculty.

# VI

This next faculty is that of thought or mind, whose activities are knowing, thinking and making judgements.

Before we proceed with Aristotle to consider in a general way the faculty of thought, let us recap schematically what we have learnt so far about the operations of the sensitive soul. (1) There are 'real things' in the world, consisting of matter-and-form conjointly. And there are sense-organs in the animal body with a potential for sensing. (2) Certain qualities (e.g. colour) of these 'things themselves' reach the appropriate sense-organ (such as the eye) through the continuous medium of air or water. (3) Thus a movement is transmitted to the sense-organ, and this is the sense-organ being actualised: it 'becomes one with' what it senses – a quality. (4) The effect of this is that an aspect of the form of the 'thing itself' has been abstracted by the sense-organ from the matter-and-form of the 'thing itself'. (5) The different qualities, these different aspects of the 'form' of the 'thing itself', are brought together from the sense-organs in the Imagination. An image is presented which consists entirely of (aspects of) the abstracted form of the 'thing itself'.

We can see from this that it is of the nature of the senses, of perception (that is, the senses + the Imagination), and also therefore of the mind, that they can and must deal, not with the 'things themselves', but with the 'form' of the 'things themselves' abstracted from their matter. However, we need constantly to remember that while the senses, perception and mind cannot but help make this abstraction, the 'things themselves' are, and remain, matter-and-form conjointly, and their form cannot and does not exist apart from their matter. What we have to remember is that the 'forms' with which the mind deals are only representations of the 'things themselves': they are not, and cannot be, the same as the 'things themselves'. We think abstractions (form-without-matter); but the world contains only 'real things' (form-and-matter conjointly). Thus it is of the nature of the mind that it can deal only with 'beings of reason' – concepts – which parallel, but are different kinds of thing from, 'real beings'. Unless we remember this we will confuse our mental *abstractions* of 'things themselves' with the 'things themselves'; we will assume that abstractions (form-without-matter) actually exist outside our minds somewhere. (This is one of the errors that Aristotle believed Plato had fallen into.)

To return to Aristotle's account of the intellective soul: the first of the activities of the faculty of thought or mind involves inspecting the image presented in Imagination, and making judgements about it. First we need to judge whether (for instance) the whiteness presented in the Imagination is or is not the whiteness of a particular 'thing itself' (of a particular man, for instance) in the world. Here is the first opportunity for error. And second, we need to make judgements about the 'common sensibles': about the movement, shape, magnitude and number of the 'thing itself'. Here again there is room for error, for being either right or wrong. The senses themselves (as we have seen) are not capable of error; neither is the Imagination when working under the direct stimulus of the senses. But the activity of judging about what is presented in the Imagination is a minefield of potential error.

This particular issue would be a matter of the greatest concern to William Harvey: how can we minimise or eliminate error in these acts of judgement? It is a matter of the greatest moment, since the whole possibility of gaining proper knowledge about the world hinges on it. For it concerns whether a particular 'real thing' in the world does, or does not, actually exist, and whether it does or does not possess certain attributes. If we can indeed eliminate the possibility of error here, then the road to knowledge about the world is open. If, however, error cannot be eliminated from these acts of judgement, then no certain knowledge of the 'things themselves' is possible. Philosophy, the search for wisdom and knowledge, becomes an impossible project.

We have seen (Aristotle reminds his reader) that the senses are acted upon by what can be sensed ('the sensible'), and that as the senses sense qualities, so they must themselves be potentially all those qualities. The same is the case with the first aspect of the mind: the Speculative Mind. The Speculative Mind must be acted upon by 'the thinkable', and it must be potentially all the things it can think. From the little summary above we have been reminded that what the mind thinks are the forms of 'real things' abstracted from their matter. The mind must therefore be potentially all the 'forms' of 'real things'. Until it thinks, the mind is only potential: it is, in Aristotle's image, like a wax tablet which bears no actual writing, yet which potentially bears all and any writing. Writing is a particularly happy metaphor for Aristotle to use, for he treats the Speculative Mind as dealing exclusively in images. Not only must the Speculative Mind be potentially all the 'forms' of 'real things', but it must itself be a 'form'. For, as with the senses, the mind when actualised (i.e. when one thinks) becomes one with its objects – and its proper objects (as we have seen) are 'forms'. Thus the Speculative Mind is a 'form' which deals in 'forms'. When he is thinking, Aristotle reports, the wise man, the philosopher, is said to 'become one with' the objects (i.e. the 'forms') that he contemplates.

In the Speculative Mind, therefore, the philosopher has a quiet grasp of the 'forms' of the 'real things' in the world: he is 'at one with' them. He knows them, knows them in the strongest, fullest, most intimate way it is possible to know anything: he knows what they are and why they are as they are. He grasps their essences. For Aristotle, the essence of some 'real thing', as grasped in the Speculative Mind, is an image, an image built up ultimately from the attributes of that 'real thing' as sensed by the senses, unified in the Imagination, and made whole by the philosopher's judgement about the 'common sensibles'. But the Speculative Mind is not 'at one with' a mere image. For, while it has to (by its nature) deal in 'forms', the Speculative Mind is 'at one with' the 'real thing' itself! For the 'form' of any 'real thing' is (as we will recall) what-makes-it-what-it-is. If you are 'at one with' what it is that makes a thing what it is, then you are 'at one with' that 'thing itself'!

## Notes

1 Pliny Secundus, *The historie of the world* (1634), pp. 200–201, spelling modernised. The original Latin, as used in the familiar Loeb edition (vol. 3 of Pliny, p. 34, tr. by H. Rackham, 1967), is as follows: Alexandro Magno rege inflammato cupidine animalium naturas noscendi delegataque hac commentatione Aristoteli, summo in omni doctrina viro, aliquot milia hominum in totius Asiae Graeciaeque tractu parere ei iussa, omnius quos venatus, aucupia piscatusque alebant quibusque vivaria, armenta, alvearia, piscinae, aviaria in cura erant, ne quid usquam genitum ignoraretur ab eo. quos percunctando quinquaginta ferme

volumina illa praeclara de animalibus condidit. quae a me collecta in artum cum iis quae ignoraverat quaes ut legentes boni consulant, in universis rerum naturae operibus medioque clarissimi regum omnium desiderio cura nostra breviter perigrinantes.

2 See for instance how D'Arcy Thompson chose to place this supposed event at the beginning of Aristotle's scholarly life: below, Chapter 9, 'Precursing Aristotle'.

3 It is not possible to put Aristotle's treatises into chronological order with any certainty. If a scholar claims a particular chronological order for them, it is to lend weight to a particular interpretation.

4 On which in general see Durrant (ed.), *Aristotle's De Anima in focus* (1993), and Corcilius, 'Soul, parts of the soul' (2021).

5 One or two other scholars have made points similar but not identical to this. See for instance Lloyd, 'Apects' (1992). Unlike me, Lloyd is not here concerned with anatomy. I remain an admirer of Frederick Woodbridge whose *Aristotle's vision of nature* (1930/1965) presents a clear, and amusing, analysis of *De anima*.

6 Cornford, *Before and After Socrates* (1979). p. 64.

7 Marjorie Grene explains their relative positions thus: 'the core of Plato's doctrine, the Forms, recollection, the dualism of soul and body, Aristotle was at pains to refute ... [Aristotle agreed with Plato that] it is forms that the knowing mind properly and rightly knows. Further, form is causal, it is the reason why things are what they are and the reason why our minds can know them. Further still, form is the source of unity: it is one as against the multiplicity of the informed, or of the unformed. Moreover, the mind in knowing form is somehow like it, at one with it. All this is common ground. But Aristotle found form, intelligibility, definiteness, where Plato had never found it: in the limited, recurrent but orderly processes of nature itself', *A Portrait of Aristotle* (1963), p.65.

8 As for instance by Gotthelf: 'With these words Aristotle introduced his students to the study of biology'; see *Aristotle on Nature* (1985), p. vii. Similarly, Lloyd writes that this chapter 'provides a fascinating insight into the resistance that Aristotle had to overcome among some of his contemporaries in order to get biology accepted as a worthy subject of the philosopher's investigations', *Aristotle* (1968), p. 71.

9 Here and elsewhere I follow here the Loeb translation by Peck, modified according to the advice of Dr Christine Salazar, for whose assistance I am most grateful.

10 Cherniss, *Aristotle's Criticism of Presocratic Philosophy* (1933).

# 2   Aristotle *On the causes of the parts of animals*

The animal books of Aristotle were written originally in Greek. But, like his other writings, they are usually known by Latin titles, since it was in this form that they were the staple texts of the universities of Western Europe for hundreds of years, where all the teaching took place in Latin. And it is the case that William Harvey knew and read them in Latin, not Greek. The main ones are these:

1. *History of animals (Historia animalium)*;
2. *On the parts of animals* – which should properly be rendered *On the causes of the parts of animals* (*De partibus animalium*);
3. *On the generation of animals* (*De generatione animalium*);
4. *On the movement of animals* (*De motu animalium*);
5. *On the progression of animals* [i.e., how they move or walk] (*De incessu animalium*);
6. *On the soul* (*De anima*)

It has been pointed out by the translator A.L. Peck, that *De anima* 'is necessary to the completeness of the scheme [of Aristotle's zoological works], but as it has given rise to a whole department of study, it is usually treated apart from the rest'.[1] This unfortunate development in the modern-day study of Aristotle has, in my view, somewhat neutered the study of Aristotle's animal books. Altogether the animal books represent about a quarter of the surviving texts by Aristotle.

In the Greek tradition, as we have already seen, the central role of the philosopher – the 'lover of wisdom' – was to seek out causes of why things in the world are as they are. This continued to be the central role of philosophers until the late 18th century in Europe. It is strikingly different from the role of the philosopher today.

We now continue our short account of Aristotle's book on the soul. No-one, Aristotle is claiming, can investigate the soul properly unless they start their enquiry from its actual manifestations in living creatures. The nature of the soul is the object of Aristotle's enquiry here: and it is this enquiry which leads him to look at living creatures. Aristotle looks at *animals* because he is interested in the *soul*. He looks at them solely with respect to

DOI: 10.4324/9781003247616-4

what light they shed on the essence and attributes of the soul. He looks at animals because they are the soul-in-action.

This is beginning to sound very like Plato's claim that the (human) body is simply the 'vehicle of the soul', and his insistence likewise that the body exists 'for the sake of' the soul. Aristotle is indeed dealing with this issue because he has inherited this way of thinking from Plato his teacher. But Aristotle is turning the issue upside down. He is saying that instead of *assuming* what kind of thing the soul is, what its requirements are and how the body must therefore be structured and function to fulfil these requirements (as in his eyes Plato has done), what the philosopher needs to do is to investigate and hence *discover* the nature of the soul from the soul-in-action. This central concept, the 'for the sake of which' will be the grand clue to enable the investigator to understand why animals are the way they are, why they have the parts they have and why those parts function the way they do. Investigating animals is the necessary means to the end of understanding 'The Animal' – which is 'The Soul in action'.

As we have seen, in order to accomplish this plan, and to gain an understanding of 'the soul in action', it was thus essential for Aristotle either to systematically cut up animals himself – that is, to anatomise them, or to use the anatomical findings of other people. The use of the knife was central to his investigations, because you cannot reach the internal organs without cutting animals open, either alive or dead, and the organs are the instruments of the soul. Hence Aristotle turned to the work which is recorded in the animal books.[2]

So: while we have established that it is, to put it mildly, an anachronism to credit Aristotle with being a biologist, he was nevertheless an anatomist. Moreover, he was an anatomist in both immediate senses, in that he used both his hands and his head to investigate the bodies of living beings, alive and dead, in order to discover reasons why they are as they are. But what kind of anatomist was he in particular? For there are many ways of undertaking anatomy, using different manual techniques and (in particular) using different concepts of exploration and explanation. When he was anatomising, what was Aristotle doing, and why?

Aristotle's main statement of his own programme of investigation into animals comes from early in his *Historia animalium*. After citing various general ways in which animals differ (some are aquatic, others are terrestrial, some are viviparous, others are oviparous, and so on) Aristotle wrote:

> So these things – about which we shall speak later with exactness – have now been said in this manner in outline as a foretaste about all such things; all must be inspected so that first we might grasp the recorded differentiae and the attributes. After this we must attempt to discover the causes of them. For the method must be followed thus in accordance with nature, since the recorded *historia* [is] about the particular: for demonstration must be about the particulars and from them; out of them it obviously arises.

> (491 a 7–14)

Aristotle then goes on to say that we should first look at the differences in the parts of animals, one animal perhaps lacking a part another has, or the parts being different in arrangement in one animal rather than another.

Aristotle is informative in a number of other places in the animal books about what his programme and his procedure are. For the sake of brevity we can here gather together some more of the programmatic statements that he makes in *Parts of Animals,* a book whose title ought properly to be rendered *Of the Causes of the Parts of Animals*. Aristotle asks: 'should the student of Nature follow the same sort of procedure as the mathematician follows in his astronomical expositions – that is to say, should he consider first of all the phenomena which occur in animals, and the parts of each of them, and having done that go on to state the reasons and the causes …?' (639 b 6). To which the answer is yes, 'we ought first to take the phenomena that are observed in each group, and then go on to state their causes' (640 a 15). 'The best way of putting the matter would be to say that *because* the essence of man is what it is, *therefore* a man has such and such parts, since there cannot be a man without them' (640 b 1). 'We have to state how the animal is characterized, i.e. what is the essence and character of the animal itself, as well as describing each of its parts; just as with the bed we have to state its Form. Now it may be that the Form of any living creature is Soul, or some part of Soul, or something that involves Soul' (641 a 17). '[The student of Nature should] inform himself concerning Soul, and treat of it in his exposition; not, perhaps, in its entirety, but of that special part of it which causes the living creature to be such as it is. He must say what Soul, or that special part of Soul is; and when he has said what its essence is, he must treat of the attributes which are attached to an essence of that character' (641 a 25). And finally, for the present, 'Now the body, like a hatchet, is an instrument; as well the whole body as each of its parts has a purpose, for the sake of which it is; and the body must therefore, of necessity, be such and such, and made of such and such materials, if that purpose is to be realized' (642 a 11, Peck's translation).

Aristotle's philosophical writings on the soul can be seen to be related in the following way:

| | |
|---|---|
| General | On the soul |
| | History of animals |
| | Parts of animals |
| | Dissections (treatise lost) |
| | Sleep and wake* |
| | Length and shortness of life* |
| | Youth and old age* |
| Vegetative soul | On food (treatise lost) |
| | On respiration* |
| | (On breath*) |
| | On the generation of animals |
| Sensitive soul | Sense and sensible objects |
| | (On things heard) |

| Motile soul | On the movement of animals |
|---|---|
| | On the progression of animals |
| Rational soul | On memory and recollection* |
| | Dreams* |

* Treatises later grouped as *Parva Naturalia*.

Where a title is given in brackets its authenticity is regarded as dubious.

With respect to the history of anatomising that we are here concerned with, we need to note in particular about Aristotle's soul project that for him the heart is far and away the most important organ in the animal body, for it is the centre of the most basic aspect of the animal soul, the 'vegetative soul' which rules in the whole of the trunk of the body; the heart is the first part to come into being in the animal, and the last to die. If the soul is localised anywhere, then it is in the heart. Aristotle also argues that the heart is the centre of all the other aspects of the animal soul as well: the sensitive, the motile, the rational. With the heart thus treated as the centre or seat of the vegetative soul, this means that Aristotle saw the blood vessels of the body as constituting only *one* system, and based on the heart; that is to say, he made no functional distinction between arteries and veins. Nor did he attribute much in the way of function to the brain: for him, thought had no anatomical location since the mind is solely a Form (uniquely a Form without matter), and hence there are no organs associated with it to be investigated. Such attitudes on his part were a source of great controversy after Aristotle's time, when a functional distinction between arteries and veins was made, and when the nerves were discovered, distributed from the brain.

We can use the argument of the eminent classicist David Balme to sum up what we have heard from Aristotle about the point of his animal books. In the first place Balme argues, Aristotle was not making a classification of animals. What then was he doing with all his apparently random groupings of animal differentiae in his discussions? His discussions (according to Balme) 'show how differentiae are essentially associated or divergent, and this is the real use Aristotle makes of them in *Parts of Animals* and *Generation of Animals*. In his arguments about causes there, he appeals largely to the inter-relationship of differentiae which appear to belong together. He seeks the significant, causal grouping of differentiae ... His method in fact is what he briefly describes at *Posterior Analytics* II, 98 a 14–19: by looking for the characteristics which are regularly associated we may detect their cause ... Aristotle's purpose in *Parts of Animals* and *Generation of Animals* is made clear: to find the 'causes' of animals' parts, and of their generation and growth. In doing so, it seems that an important part of his method is to look for significant differentiae and combinations of differentiae; he constantly groups and regroups them to focus on particular problems'.

And Balme says, in what might look like an incoherent jumble of observations in *History of Animals* too, Aristotle 'does state his purpose: "first to grasp the differentiae and attributes that belong to all animals; then to discover their causes" (HA I 491 a 9)'.[3]

We have to recognise that any explanation that Aristotle offered had to be true of all the apparent exceptions too: the explanation (or 'cause') of the presence and function of a given part or organ in one creature had to be one which also covered its possible different incidence, or even its total absence, in another creature. What Aristotle does in practice is to specify either what other part or organ fulfils an analogous role in the other creature, or what particular features of that animal make the part or organ in question different in its incidence, or even unnecessary. The reasons for such particular occurrences or absences lie, for Aristotle, in one or more of four aspects of the creature in question (HA 487 a 11):

1. Its life
2. Its activities
3. Its habits
4. Its (other) parts

We shall now see this programme of investigation and anatomising of animals being taken up, after centuries of neglect, by a celebrated professor of anatomy in Padua in Italy: Girolamo Fabrici, more usually known as Hieronymus Fabricius ab Aquapendente.

## Notes

1 Preface to Peck's translation of *Generation of Animals* (1942), p. vii.
2 This claim does not have any consequences for establishing the chronological order in which Aristotle went about his work; it is a claim about the logic of his argument only.
3 Balme, 'Aristotle's use of differentiae', pp. 190–192.

# 3 Aristotle's animal in Padua

## The anatomical investigations of Fabricius[1]

### Fair Padua and Aristotle's philosophy

In 1611 the English traveller Thomas Coryat published his *Crudities* or his 'silly Observations' (Epistle Dedicatorie) to encourage courtly gentlemen to travel overseas and seek learning. A gentleman might see 'flourishing Universities ... furnish'd with store of learned men of all faculties, by whose conversation a learned traveller may much informe and augment his knowledge' (Epistle to the reader). In five months of travel in the middle of 1608, Coryat had visited seven countries, and taken enough notes to fill two large printed volumes. It was all quite a rush, and in the three days he spent in Padua he hardly had time to see the university, even though he knew it was very important:

> Truely I must needs lay an imputation of great indiscretion upon my selfe, in that being in so famous a University as this I omitted to see their Colledges, which are in number nine, heare their exercises and disputations, observe their statutes and priviledges, the foundations and revenues of their houses, discourse with some of their learned men & professors, and note such other worthy things as are observable in so noble an Academy. For my minde was so drawe away with the pleasure of other rarities and antiquities, that I neglected that which indeed was the principalest of all.
>
> (p. 153)

He was so keen to copy down inscriptions that he had to rely on a couple of English students for his information about the university itself:

> I heard that when the number of the Students is full, there are at the least one thousand five hundred here: the principall faculties that are professed in the University, being physicke and the civill law: and more students of forraine and remote nations doe live in Padua, then in any one University of Christendom. For hither come in, many from France, high Germany, the Netherlands, england, &c., who with great desire flocke together to

DOI: 10.4324/9781003247616-5

Padua for good letters sake, as to a fertile nursery, and sweete emporium and mart town of learning.

(pp. 154–155)

The comparison of the university of Padua to a nursery of learning appears also in Shakespeare's play *The Taming of the Shrew*, written probably just a little earlier, in the early 1590s. Here two visitors, a young master and his servant, are discussing the relative merits of the educational offerings of Pisa and Padua universities – though it turns out in the play that, typically for male students, they are actually more interested in meeting girls than in pursuing philosophy. Lucentio the master speaks to his servant Tranio:

> Tranio, since for the great desire I had
> To see fair Padua, nursery of arts,
> I am arrived for fruitful Lombardy,
> The pleasant garden of great Italy ...
> Here let us breathe, and haply institute
> A course of learning and ingenious studies ...
> Tell me thy mind, for I have Pisa left
> And am to Padua come as he that leaves
> A shallow plash to plunge him in the deep,
> And with satiety seeks to quench his thirst.

Apart from locating Padua in Lombardy, rather than in the Veneto, Shakespeare here well illustrates the attraction of Padua university at the end of the 16th century: for some subjects – and especially medicine – it was simply the best university in the world, and worth trekking hundreds of miles over several weeks to attend. A medical degree from Padua had, to put it mildly, a certain cachet that no other university could offer. And one reason for this is that at Padua professors were actively encouraged to undertake new investigations in their subject, to teach the latest thing and to question received opinion. And for this to be the case it meant that Padua was the university most free from the interference of the Catholic Church – an independence that was guaranteed them by their masters, the merchant Senators of Venice.

The 16th-century return to Aristotle the anatomist was of a piece with a particular larger programme of return to Aristotle the whole philosopher at Padua, a programme which produced major novelties in the understanding of what Aristotle was saying and what philosophy was about. Aristotle had of course been the mainstay of the curriculum since the 13th century when universities began, with pretty much the same selection of his books being read and commented on over the intervening centuries. Modern studies are revealing that the 16th century was perhaps the greatest period of flourishing of the study of Aristotle – whereas historians had hitherto assumed that Aristotle's texts were dropped in the 16th century in favour of anti- or non-Aristotelian approaches. Every educated man knew his Aristotle.

The particular revival of Aristotle with which we are here concerned, which involved a general programme of resuscitating the works of Aristotle together with the revival of his anatomical project, was unique to Padua, the home university of the republic of Venice, and it was promoted by the committee of Riformatori who ran the university for the Venetian senate. We may start our view of teaching at Padua by casting our eyes over one of the printed announcements of lectures, taking for convenience the one for the session beginning November 1593 as our example.[2] There are 26 professorships listed, including one vacancy. On the original document in addition to the names of the professors, the hours of lecturing and the texts are also specified for each position. 'Ordinary' professors are senior to 'extra-ordinary' ones; the chairs 'in the way of St Thomas' (St Thomas Aquinas) and 'in the way of Scotus' (Duns Scotus) were usually occupied by Dominicans and Franciscans respectively. The list is in the published order and indicates the hierarchy of the chairs and subjects:

> Chair in Theology in the way of St Thomas
> Chair in Theology in the way of Scotus
> Chair in Holy Scripture
> Chair in Metaphysics in the way of St Thomas} *Metaphysics* Book 1
> Chair in Metaphysics in the way of Scotus} *Metaphysics* Book 1
> Two Ordinary Chairs in the Theory of Medicine
> Two Ordinary Chairs in the Practice of Medicine
> Two Ordinary Chairs of Philosophy} *On the Soul* Books 1 & 2
> Two Extraordinary Chairs in the Theory of Medicine
> Two Extraordinary Chairs in the Practice of Medicine
> Two Extraordinary Chairs of Philosophy} *On the Soul* Book 3
> Chair in the Moral Philosophy of Aristotle (vacant)
> Chair in Surgery and Anatomy
> Chair in the Third Book of Avicenna
> Chair in Simples (i.e. plants for medicines)
> Three Chairs in Logic
> Chair in Mathematics
> Chair in Greek and Latin Humanitas

To a modern eye, the amount of Aristotle being read is quite astonishing: the two professors of metaphysics both read the same work by Aristotle, *Metaphysics*; the two ordinary and the two extraordinary professors of philosophy all read Aristotle's *On the Soul*; there is a professor of the moral philosophy of Aristotle, who would normally read Aristotle's *Ethics*; and the three professors of logic read the first three books of Aristotle's *Posterior Analytics*. This teaching list of course reveals only part of the complete cycle of texts to be read by the professors. For instance the metaphysics professors had to read not only book 1 but also books 7 and 12 of *Metaphysics*. Similarly, the ordinary and extraordinary professors of philosophy had to read books 1, 2 and 8 of Aristotle's *Physics*, the two books of *On Generation and*

*Corruption*, the three books of *On the Soul*, and the four *On the Heaven and the Earth* in the course of the full cycle of lectures, while only their reading of *On the Soul* is reflected here in this list.

While this is fundamentally still a 14th-century curriculum being taught in the late 16th century, it had received some modifications in the interim, though the specified texts of Aristotle had not been changed. The two chairs of theology and the two of metaphysics were a response (common to many universities) to late 15th-century demands for rival interpretations by the rival orders of friar; the chair in Holy Scripture had originally been a chair in the *Sentences* of Peter Lombard, and was in 1551 updated to be on the texts of the *Old and New Testaments*. The chair in mathematics (held in 1593 by Galileo) had been established in 1517 to teach 'astronomy and mathematics', in other words astrology for medicine. At Padua most of the professorships had been established in pairs, so that each had an antagonist or *concurrens* professor of different views who lectured at the same hours and on the same text; this is the case wherever in the list there are two chairs of the same subject and status. This unusual system promoted a great deal of healthy controversy between colleagues, which often spilled over into print.[3]

We can see that, as had been traditional at Padua, medicine and philosophy (arts) are still studied and taught in one joint faculty and that Paduan medical students were expected to study philosophy with their medicine; that through the system of concurrent professorships this philosophy was deliberately multifarious in interpretation but that it exclusively discussed Aristotle and the issues raised by his writings; that it was the philosophy professors who taught the 'nature' books of Aristotle; and that Aristotle's book on the soul, *De Anima*, played a major role in the philosophy teaching of those same professors. To put it another way, the question of the soul – Aristotle's view of the soul – was basic to the teaching of Natural Philosophy, and that a renewed Aristotelian Natural Philosophy was, in turn, fundamental at Padua to both medicine and anatomy. To the significance of this we now turn.

The Paduan philosophers were concerned with trying to reach Aristotle's own sentiment. One of the most famous of them, for instance, giving an oration to the students 'In praise of the study of philosophy' at Padua in 1585, said

> I will arrange what I am going to say according to two heads: for since our aim is to be skilled in the philosophy of Aristotle, we must see in the first place what mode of philosophising Aristotle himself used, and then find what we need to do so that we may be said to philosophise correctly and 'aristotelically' on Aristotle.[4]

Randall has traced the way in which discussion of the soul, of its capacity to acquire knowledge and of the mode of its existence, was conducted at Padua in a progressively more 'aristotelical' mode in the course of the 16th century, and he has indicated how it came to be divested of Platonic and Averroist interpretations by giving privileged attention to the ancient Greek commentary of

Alexander of Aphrodisias.[5] Beginning with Petrus Pomponazzi in the early decades of the century, this culminated in the period 1564–1589, when Jacobus Zabarella was one of the professors of philosophy. In Zabarella, Randall claims,

> Aristotle's thought has been made so transparently his own that it seems unjust to call him a follower: it is Aristotle himself, speaking in the Latin of Padua – not the syllogistic Aristotle with a category for every emergency – but the Aristotle who insists on the primacy of subject-matter, of fact, of experience'.[6]

Zabarella's time as Paduan professor coincided with that of Caesar Cremonini (professor 1573–1631), and Franciscus Piccolomini (professor 1560–1607), who also had reputations as being new Aristotles – it being said of Cremonini, for instance, that he was 'the Genius of Aristotle' – though they did not necessarily agree with each other in every detail of their understanding of Aristotle.

In general these Paduan Aristotelian philosophers claimed that the soul is the act of the body. As Zabarella put it:

> The body is not the act of the soul, but the soul is the act of the body; it is not body, but something belonging to body, namely its act and perfection, whence the soul cannot exist without the body, since a perfection cannot exist without that of which it is the perfection ... He who does not see that Aristotle plainly says that every soul is an informing form, is blind.[7]

If the soul cannot exist without the body, then the implication is clear: that when the body dies the soul dies; or, rather, the death of the body is the death of the soul. The soul is mortal. And this is contrary to Roman Catholic and Protestant teaching about the immortality of the soul. And if the soul indeed cannot exist without the body, then a second implication is also clear: that the soul can be understood only through its operations in the body. If we want to understand what makes a man a man, or a plant a plant – that is, their soul, their informing form – then we can only do so by studying the operations of their respective bodies, for the soul 'is the act of the body'.

## Fabricius, Professor of Surgery and of Anatomy

The man who resurrected Aristotle's research programme in anatomy was the Paduan Professor of Surgery and Anatomy, Hieronymus Fabricius ab Aquapendente (Girolamo Fabrici or Fabrizie), whom I shall be referring to by his Latinized name as 'Fabricius'.[8] It was Fabricius who introduced this new Aristotelian anatomy at Padua, and who first read and practised Aristotle 'aristotelically' in anatomy. He held the post of Professor of Surgery and lecturer in Anatomy from 1565, when he was about 32 years old, until 1613, when he was about 80, the last dozen or so years with the title 'Professor

Superordinarius'. His winter months were dedicated to anatomy, his summers to surgery.[9]

In fact Fabricius had three jobs to perform. The first was the surgery teaching, which he took very seriously. Toward the end of his teaching life there appeared at Frankfurt in 1604 his *Pentateuchos Cheirurgicum* as 'presented by the Author in Padua University in his public lectures'. This refers to the first five books (as in the Bible), covering the five kinds of external accidents that the surgeon had to deal with: tumours, wounds, ulcers, fractures, and luxations or dislocations. It was produced by Johannes Hartmann Beyer, a Frankfurt doctor who had studied under Fabricius, but had not sought Fabricius' consent to publish.

Fabricius was sufficiently displeased by this to now turn to publishing his *Opera Chirurgica* himself, and it appeared in Venice in 1619. In his Dedication of the volume to King Sigismund III, Fabricius describes the contents as 'works from extreme old age, barely completed', but perfected over 50 years 'in which I actively taught Surgery and Anatomy in this most flourishing Paduan Academy, and practised every branch of medicine with energy, experiment (*experimento*), industry, diligence, as much as I was able'.

The first part of the volume written and published by Fabricius deals with all the customary surgical operations which are carried out on the human body from head to toe, Fabricius calling it 'a work perfected with great diligence and long experience'. In his Dedication dated January 1618 Fabricius writes (quoting Virgil, Ecologue 4) that 'If the latter end of a long life yet be mine' then he hopes to issue coloured images of the shapes of the surgical instruments and of the external diseases pertaining to this part of medicine. It seems that his hope was misplaced (he was after all over 80 years old) and the book was later published sometimes with and sometimes without illustrations of surgical tools, some of which Fabricius had himself invented. He speaks of the masses of instruments both ancient and modern and some that he has invented, saying 'it should be no surprise that someone like me, practising in this area of medicine for over fifty years, has prepared a most beautiful workshop (*officina*) for myself of all the instruments which are necessary for the surgeon'.[10] The text is full of descriptions of instruments of his own invention. Similarly it is full of methods of his own invention, sometimes with examples of patients treated in his Paduan and Venetian surgical practice. The ancient authorities in surgery that Fabricius discusses include (of course) Hippocrates and Galen, and also Celsus, Paulus, Rhazes, Albucasis and others. Among the more modern surgeons he gives special attention to Guido and Fallopius. His treatment of the surgical operations is from head to toe. Some tantalising moments from his teaching of surgery have been brought to light recently from a German student's notes.[11]

As the second part of the volume Fabricius puts out the *Pentateuchos Cheirurgicum* which had first been published in Germany, but 'amended in many places by the author himself'.

This great two-part *Opera Chirurgica*, a monument to a lifetime of devoted and experimental surgical practice, was issued repeatedly by publishers and editors across western Europe for almost a century. This indicates something of the esteem in which Fabricius' surgical teaching was held.[12]

It must be borne in mind that in the 16th century, the time of Fabricius, Galen was being revealed by new editions and translations of his work not only as the most important *medical* authority of Antiquity, but also the most important *surgical* authority too.

It is well known that Galen was also the most important authority of Antiquity for *human anatomy*. It was during Fabricius' reign as Professor of Surgery and Anatomy that the fabulous Paduan anatomical theatre was built (completed 1594) in the new university buildings.[13] Fabricius is customarily credited with first discovering the so-called 'valves in the veins', though he did not think of their function as 'valves', but rather as structures which delayed the outward flow of blood in the veins. It would be William Harvey who first saw them as valves in the modern sense, preventing back-flow of blood in the veins.

From Fabricius' earliest years in post, his *anatomical* engagement was two-fold. First there was the obligation to perform *human* dissection. For this each year the body of a man and a woman were supposed to be supplied by the Paduan authorities, sometimes criminals who had been put to death; but sometimes there were fewer or none. These were the official public dissections, but there were often private dissections offered at Fabricius' home. The ex-pupil of his who first published Fabricius' surgical lectures said that by the very early 1600s 'he established the anatomising of the human body for over thirty years publicly every year with dissected corpses', and we know that he both dissected and showed the parts himself.

Though he held his dual chair in surgery and anatomy for so many years, Fabricius had an uneasy relation with the students who came to Padua to learn anatomy. Many objections were made against him by generations of students for failing to perform his duties and for teaching the wrong things.[14] Yet Fabricius was nevertheless maintained and promoted in his post by the Riformatori, who kept increasing his salary. As the presence of the students was what made the university a successful business for Venice, the continued support of Fabricius by the Riformatori indicates very strong commitment on their part to the content and nature of his teaching and their conviction that what he taught was what the students really needed, even if some of them occasionally did not appreciate it.

And this brings us to the other part of Fabricius' anatomical work, which was *research*, something very much encouraged at Padua.[15] And this research was Aristotelian rather than Galenic in its form and purpose. And this may explain some of the student dissatisfaction with his anatomical teaching: they came to hear one thing, Fabricius teaches them something else.[16]

Reading Aristotle 'aristotelically' on anatomy, Fabricius of course read from *De Anima* (*On the Soul*) outwards, and hence his anatomising dealt with the different faculties of the soul, and his object of inquiry was 'The Animal'. Like Aristotle, he looked at the body as the soul-in-action; or, in the words of his own colleague for many years, Zabarella: 'the soul is the act of the body'. Fabricius' anatomical work was thus primarily about the soul, and about the body itself only in a secondary sense. Thus what Fabricius was studying was the soul, and he was teaching it to students who had already had the opportunity of acquiring extensive exposure to Paduan ways of understanding Aristotle on the soul.

Some of Fabricius' Aristotelian anatomical research and teaching is recorded, in a refined and revised form, in his other great written work. This is a great book which he set about publishing in his old age, made up of part-numbers, many but not all of which he succeeded in finishing and publishing before his death, and it was called *Theatrum totius animalis fabricae*, 'A Theatre of the Whole Animal Fabric' (1600–1621). The theme, it will be noted right from the title, is not the Galenic one of 'man' but the Aristotelian one of 'The Animal', and Fabricius' usage of the term 'fabric' was much wider than that of Vesalius. As in the case of Aristotle himself this does not mean that Fabricius was engaged in 'comparative anatomy', nor in 'biology', but in that now-lost study, the study of 'The Animal' and its 'œconomy', a discipline which for Fabricius was part of philosophy. Man was treated by Fabricius as one kind of animal, but only one.

These are the works which went toward making up *The Theatre of the Whole Animal Fabric*:

1. On vision, or on the eye the organ of sight, 1600
2. On the voice, or on the larynx the organ of the voice, 1600
3. On hearing, or on the ear the organ of hearing, 1600
4. On the formed fetus, 1600
5. On speech and its instruments, 1603
6. On the vocal communication of brutes, 1603
7. On the ostiola in the veins, 1603
8. On the artifice of the muscle, 1614
9. On the articulations of the bones, 1614
10. On respiration and its instruments, 1615
11. On the local motion of animals as a whole, 1618
12. On the gullet, stomach, intestines, 1618
13. On the covering of the whole animal [i.e. on the skin, etc], 1618
14. On the formation of the egg and of the chick, 1621 (posth.)
15. On the semen (planned but not completed).

To illustrate the extent to which Fabricius' project is identified with that of Aristotle, we may tabulate the contents of his *Theatre* against the soul anatomy of Aristotle (see above, Chapter 1):

| *Aristotelian aspects of soul* | *Works by Fabricius (parts of the Theatre)* |
| --- | --- |
| Vegetative Soul | On respiration 10.<br>On the generation of animals:<br>    On the semen 15.<br>    On the formation of the egg and<br>      of the chick 14.<br>    On the formed fetus 4.<br>On the *ostiola* in the veins 7. |

| | |
|---|---|
| Sensitive Soul | On vision 1. |
| | On hearing 3. |
| | On the covering of the whole animal 13. |
| Motile Soul | On the artifice of muscle 8. |
| | On the articulations of the bones 9. |
| | On the local motion of animals 11. |
| Rational Soul | On the voice 2. |
| | On speech and its instruments 5. |
| | On the vocal communication of brutes 6. |

The coherence of Fabricius' works in and with the Aristotelian scheme of anatomical things is clear from this table. Such topics had simply not been pursued by previous anatomists in this way, since they are Aristotelian through and through. Fabricius' commitment to this specifically Aristotelian project in anatomy is not something that he justifies at any length in his published writings, but after being engaged in successfully teaching and researching on it for some 35 years before publishing, and having outlived his rivals, he may no longer have felt any need to. Fabricius regarded his task as that of going further in the direction indicated by Aristotle: he did not think that Aristotle had said it all, and that his own role should consist only in teaching the established truths of Aristotle.

In his *De Methodis,* his colleague the philosopher Zabarella expressed this sentiment thus:

> Although I wish to be second to no mortal in admiration of the great genius of Aristotle, still I believe that he neither could have written everything or known everything, nor did he so follow the truth in all that he did write that he was never able to err; for he was a man, and not God.[17]

And here is Fabricius himself talking similarly about how he, in anatomy, is equally dedicated to Aristotle, and yet is also intent on continuing the anatomical inquiry beyond Aristotle. He is writing about the generation of animals:

> few of the Ancients and none at all of the Moderns have addressed themselves to this topic. Why this has happened I do not know, since indeed it is unworthy that such great marvels of nature should lie hidden from us. We shall reveal them with as much brevity as we can, and so effect it both by the placing of illustrations and by the plan of exposition, that anyone henceforth may be able by himself to understand and contemplate those first beginnings of the life of every animal. In this way we shall both follow and expound that great interpreter of nature, Aristotle, who first and alone enquired into those mysteries; and if anything at times escaped him, we shall point it out.[18]

Only Aristotle, that great interpreter of Nature, has pursued this topic of the generation of animals before; we shall add to his work 'if anything at times escaped him'. Fabricius made similar claims about the study of the local motion of animals and about respiration: that he was the first to resume the study of these subjects since Aristotle.

It is evident from the table above that Fabricius found topics for inquiry by simply following Aristotle: he investigates 'The Animal' as soul-in-action, and especially the vegetative, sensitive and motile aspects of the soul. In practice that means he concentrates on investigating *processes*, such as generation and respiration, and the organs involved in carrying them out, and in all animals. This is, as it had been for Aristotle, anatomy as philosophy. There is no room in this inquiry for the traditional 'three venter' division of the body, nor for the skeleton-muscles-arteries-veins-nerves sequence of Vesalius, nor for the 'three rivers' approach of Columbus, all of which approaches had anyway taken man as their central object of study.[19] Fabricius' new object of study meant new things became visible, because new questions were being asked. It was an open-ended research programme. The capacity of this approach to find anatomical novelty – including, in time and quite unexpectedly, the circulation of the blood – makes it desirable that we give an outline of Fabricius' actual practice of anatomising here.

A full account of any organ or part[20] involved, for Fabricius, a four-fold presentation[21]:

1.   the *historia*, or description of the structure
2.   an enquiry into the *action* of the part
3.   a specification of the *uses*/usefulnesses of the part
4.   a demonstration of the truth and applicability of his findings.

Only such an account was properly 'philosophical' by Fabricius' standards. Fabricius was more than once critical of Vesalius for having performed only the first of these, and hence for not having given a properly philosophical account of anatomy. Fabricius' employment of the fourth stage, demonstration, may well owe something to the discussions of his Paduan philosopher colleagues, and their development of the 'regressus' technique in this period.[22]

To produce one of his philosophical anatomical accounts, therefore, Fabricius went through a particular routine of investigation. We must bear in mind that Fabricius is always investigating some operation of the soul, such as nutrition, generation, sensation or motion, and then looking at the organs which carry it out. Thus his goal in any particular anatomical investigation was a single, general, universal account of the operation in question in that single, general, universal creature 'The Animal', and the organ(s) or part(s) which carry it out, for this would capture the essence of that particular instrument of the soul, and hence explain what it was and why it was as it was.

(1)   First he had to create the *historia*. This involved him looking at as many instances as possible of the part in question in as many kinds of animal as he could. Thus if, for instance, he was investigating respiration, and

hence looking at the lung, he would investigate the incidence of the lung in all the creatures that have a lung, and also he would note that the lung was absent in certain creatures (such as fish and insects), and note what other organ (if any) seemed to be standing in place of it. This involved him in asking a regular series of questions every time; these questions were based on the 'categories' of Aristotle. He asked: is it present? How many are present? What form or shape does it have? What position is it in? What connections does it have? and so on.

(1a) Fabricius then took this data from the series of separate incidences, and looked for the similarities in the answers to each of these questions. Hence he could make an induction and reach the general incidence. He could now say (for instance) that in the accomplishment of (say) vision, it is usually the case – it is the norm – that part x does occur in animals, it is usually of shape y, it is in position z, it is connected to other parts a and b, and so on. This is the general *historia*. When he came to write-up this *historia* for publication, Fabricius would now write in a general, universal way, talking for instance of 'the eye' rather than of the eyes of particular animals. He would thus have reached the general through a study of the particular.

(1b) At the stage of assessing the similarities, Fabricius would also need to make a listing and grouping of the dissimilarities in the answers to his set of categorical questions, because his final account would need to be able to explain these as well, would need to be able to account for why in some animals there is no eye, why it is in some animals in an unusual position, why it is structured in an unusual way, and so on. For, unless the general, universal, account allowed one to answer these questions satisfactorily, then it was not a proper general, universal, account. Ultimately he would be able to give answers about these dissimilarities in terms of the particular requirements or peculiarities of the life, activities, habits or other parts of the particular animal. This is just like Aristotle's procedure of course.

(2) Then Fabricius turned to investigating the particular *action* of the part in question. An 'action' in this sense is some unique and public role contributing to and indispensable to the effective functioning of the whole animal, such as seeing, hearing, making blood, or whatever. As a good Aristotelian, Fabricius would assume that the capacity or faculty of a particular part to perform a particular action usually resided in the special nature of the material of that part: thus the liver has a blood-making faculty, the heart a pulsatile faculty. So in identifying and localising the 'action', Fabricius would be looking for the special and unique kind of material constituting the part, in whose substance the unique action was localised; for instance he discovered that the crystalline humor of the eye is where the unique action of the eye, seeing, takes place.

(3)  In the next place Fabricius would try and discover the *use* or cause of the part in question: the general account of why it exists and why it is as it is in The Animal. For instance, of the placenta the 'first and chief useful-ness', Fabricius discovers, is 'the protection and defence' of the uterine vessels.

(4)  Finally Fabricius would give a demonstration of the truth of his account. Having ascended from the particulars to the general by induc-tion (from the animals to The Animal), such 'demonstration' could be used to descend from the general to the particulars, and thus show how the particulars are comprehended in – are instances or instantiations of – the general. In the case of anatomy such a 'demonstration' shows how the incidence of a given part as it occurs in each and every instance of animal that actually exists in nature accords with, and is an instance of, the general account of it in The Animal. But in practice Fabricius tended to use this 'demonstration' only to deal with particular varia-tions from that norm in particular animals. What he therefore usually demonstrates is how the particular incidence (or absence) of the part in certain animals suits, is necessary for, and is explained by, their particu-lar form of life, or activities, or habits or other parts. That is to say, he shows how they are indeed particular instances of the general case, and not exceptions to it.

We can illustrate this way of proceeding[23] by looking at Fabricius' most famous discovery, the *ostiola*, what historians have called the 'valves in the veins'. Fabricius claimed to have discovered these in the human around the year 1574 (we recall that he was of course obliged to demonstrate human anatomy to the surgical and medical students). What he had found were little membranes which occur only in the veins, and only in some of the veins. There are no such membranes in the arteries even though the function and structure of the arteries is comparable to that of the veins. So Fabricius has to be able to explain – and his research will not be finished until he can satisfactorily explain – why these membranes exist in some veins but not all veins, why they do not exist in the arteries (why they are not needed), what their action is, what role or 'use' they fulfil when present in the œconomy of The Animal and what are the conditions that make them unnecessary when they are absent. We cannot, of course, read Fabricius' published account as if it were a step-by-step account of his research, like a laboratory note-book; it presents only the results of his enquiry, and in a polished form. But if we bear this in mind, a great deal about his enquiry can be seen from simply looking at the skeleton of the argument of his little book. Fabricius is of course tak-ing it for granted that the veins are for the distribution of nutritive blood to all parts of the body, and the arteries for the distribution of 'vital' blood.

Definition: What I am referring to by this name *'ostiola'*: delicate mem-branes in the cavity of veins, occurring singly or in pairs, mostly in the limb veins; opening upwards, and having a form like a node in a twig.

General purpose or 'use': To delay the blood, in the interest of the proper distribution and assimilation of nourishment throughout the body.
The *ostiola* are necessary in order

(a) to ensure that the 'upper' limbs receive adequate nourishment;
(b) to prevent permanent swelling in the extreme ends of limbs.

No anatomist has discovered them hitherto. Why not? Maybe because

(a) since the veins are intended for the free flow of blood, anatomists would not expect to find membranes in them;
(b) they do not occur in arteries;
(c) they do not occur in all veins (for instance the vena cava, the jugulars, the 'outer' veins).

Why arteries do not require them:

(a) arteries have a different role: although they contain blood they are not concerned with nutrition;
(b) arteries have a different structure: their thick walls are unlikely to suffer distension;
(c) arterial blood has a different movement: in arteries there is constant flux and reflux, rather than predominantly one-way flow;
(d) Why was it necessary to retard the flow of blood in the veins? In order to ensure appropriate delay for aliment to be assimilated by the parts.

Why small veins do not need them:

(a) they contain only a small amount of blood and all that suffices for them for purposes of local nutrition;
(b) their needs are met by the action of the *ostiola* in the larger veins;

Another need for the *ostiola* in the limbs (where they mostly occur):

the frequent local motion which is characteristic of the limbs creates local heat; without the existence of the *ostiola* this would naturally draw more blood to the limbs, hence creating undernourishment of the principal parts and rupture of the limb veins;

'either of which', Fabricius continues,

was going to be very pernicious to the whole animal (*toti animali*), given that it was essential that the principal parts such as the liver, heart, lungs and brain should always abound most copiously with blood. It was for this reason, I believe, that the vena cava (where it passes through the trunk of the body) and similarly the jugulars, should be quite destitute of *osti-ola*. For it was requisite that the brain, heart, lungs, liver and kidneys – which procure the conservation of the whole animal (*totius animalis*) – should abound with nourishment, and it was essential that it should not be detained in them even for a moment, both in the interests of replacing lost substance, and of producing the vital and animal spirits whereby life is conserved for animals (*animalibus*).

But if you observe *ostiola* at the beginning of the jugular veins in man, you may say that they have been placed there to detain the blood, so that in the declined position of the head it should not flood into the brain like a river and be accumulated there more than is appropriate.[24]

We can see from this that Fabricius deals with the *ostiola* as phenomena of the whole animal (*totius animalis*), giving a *historia* of them (number, form, site, distance and so on), an account of their *action*, and of their *use*. At all points his account is a general one. In the last paragraph quoted, for instance, Fabricius gives the general case on the incidence of the *ostiola* in the jugular veins, based on observation: that they are devoid of *ostiola*; and he gives the general reason for it in the life of The Animal, that here the blood must suffer no delay since the brain, as one of the principal organs, needs an unchecked supply of fresh nourishment. He then deals with an apparent exception, the case of man, where *ostiola* may be found in the jugulars. This apparent exception is resolved by relating it to the particular characteristics of the life of this particular animal. The general reason for the presence of the *ostiola* (even here) still holds good: to delay the blood. But man usually carries his head upright, and the supply of nourishment to his brain is suitably catered for in that position. So, when man bends his head, a rush of blood would occur to the brain: hence the (exceptional) presence of *ostiola* in the jugulars of man, fulfilling a special need for the life of this particular animal.

Thus even in a treatise clearly first inspired by his having discovered the *ostiola* in a human subject, and a treatise illustrated exclusively by pictures of the *ostiola* in man, Fabricius is still dealing with the incidence of the *ostiola* in all animals – and hence in The Animal – of which man is but one instance. Finally we can see why Fabricius called these membranes *ostiola*, 'little doors' (rather than 'valves', as this term is usually translated nowadays), for this is how he saw their use or function. When the part is properly identified in this way, then the enquiry is well and truly finished, for the part can now be named from its 'use' or function.

Whenever he published his final results of a piece of anatomical research Fabricius was always dealing with the general, but he had a choice about the precise form in which to present it to his reader. He could, if he chose, present the general account with relation to one particular instance, pointing out the ways in which the chosen example itself differed from the norm or, where this one substantially represented the norm, the ways in which certain other animals differed from this norm in the part or process in question. *On the Formation of the Egg and of the Chick* is an instance of his choosing this form of exposition, with the chicken as the example; so too is *On the Little Doors in the Veins*, where he uses man as his example. Alternatively, he could discuss the norm about a given process or part, and use illustrations of whichever animals most clearly or conveniently showed particular points. This is what he does for example in *On the Formed Fetus*, whose illustrations therefore show a wide range of animals. But still of course this is not 'comparative anatomy' he is engaged on, but always the anatomy of 'The Animal'.

Fabricius' concern with 'The Animal' rather than with just man, is further shown by the breadth of the range of illustrations that he used. He wanted the *Theatre* to have more than 300 illustrations, with the parts represented in their natural size 'and, what is no less important, in their natural colour', and in the interests of clarity to have each picture in two forms, one coloured, one not.[25] The coloured ones would presumably have been made under Fabricius' direction by people hand-colouring the monochrome prints at the printer's shop. Some of the published works, such as those on generation, have their full complement of (monochrome) illustrations. But some of them were never published. Many of the original coloured paintings that Fabricius prepared are now in the library of his homeland, St Mark's Library in Venice.[26]

Fabricius was the first but not the only person to pursue the anatomical project of Aristotle. He taught generations of students to see anatomy in this way, and he influenced a few of these sufficiently that they took up such anatomising themselves. Fabricius' most famous pupil was William Harvey, but he also had a successor in the Paduan surgical and anatomical chair who worked and taught along the same lines. This was Julius Casserius (Professor at Padua 1609-16) who, to judge by his printed works, certainly seems to have conducted such Aristotelian anatomy. In his Dedication to his *Anatomical History of the Organs of the Voice and Hearing* (1600–1601), he writes, 'If any contemplation of Nature is worth admiration, those ones are especially so which pertain to the faculties of the soul and their instruments'.[27] Similarly his *Pentaestheseion, That is a Book on the Five Senses*, Venice, 1609 (reissued as *Nova Anatomia*, Frankfurt, 1622), was an Aristotelian piece of anatomy. In true Aristotelian mode, Casserius is concerned with 'the sensed' as well as with the sense organ itself. His successor as Paduan anatomy teacher, Adrianus Spigelius, however, is a Vesalius-style anatomist in his *Ten books on the Fabric of the Human Body* (Frankfurt, 1632, posth.), and even in his *Book on the Formed Fetus* (Padua, 1626, posth.), which deals entirely with the human foetus. Spigelius was from Brussels (like Vesalius) and though Padua-educated had spent many years in the north of Europe before becoming Professor at Padua.

## Notes

1  This is a revised and extended version of Chapter 6 of my *Anatomical Renaissance* (1997), where fuller references can be found. My general thesis here about how the philosophy professors and their commentaries on Aristotle's *De Anima* influenced Fabricius' conception of anatomy, has more recently been confirmed and developed by De Angelis, 'From text to the body' (2008).

2  This list is at Padua university; it is published in Rossetti, *The University of Padua:* (1983), p. 35.

3  Randall, *The Career of Philosophy* (1962), p. 74.

4  'Hac de re [viz., de norma et ratione philosophandi] verba facturus ad duo capita totam orationem meam redigam; cum enim versandum nobis sit in Aristotelis philosophia, videndum primo loco est quam philosophandi rationem servaverit Aristoteles, deinde vero quid nobis agendum sit ut in Aristotele dicamur rite et aristotelice philosophari', Pra, 'Una "Oratio" Programmatica', p. 287, my translation.

5  Randall, *The Career of Philosophy* (1962), Chapter 2.

6  Randall, *The Career of Philosophy* (1962), p. 84.

7  Zabarella, *Commentarii in De Anima*, 1606, as translated and cited by Randall, *The Career of Philosophy* (1962), p. 85.

8  For Fabricius' appointment and reappointments, and his stipend and his stipend increases, see Tosoni, *Memoria* (1844), pp. 97–107.

9  For details of Fabricius' life and university roles I am indebted to Adelmann, *Embryological Treatises* (1942).

10  Dedication to 1628 edition, p. 4.

11  Stolberg, 'Learning anatomy in late sixteenth-century Padua' (2018).

12  It was subsequently issued in Latin at Frankfurt 1620, Lyons 1628, Padua 1641, Padua 1647, Padua 1666, Leyden, 1723. A Dutch version appeared at Rotterdam 1661; a French version at Lyons 1666, Lyons 1670, Lyons 1674. An Italian version Padua 1672, Bologna 1678, Padua 1685, Padua 1711. A Spanish version Madrid 1676. Quite a lot of it was translated into English and appeared in the *Bibliotheca anatomica, medica, chirurgica. &c,* 1711–1714.

13  Semenzato, *The Anatomy Theatre [at Padua]* (1995).

14  See for instance Favaro, *Atti della Nazione Germanica* (1911), vol. 1, p. 286 for an instance in 1590; and for a general consideration of these conflicts, see Adelman, *Embryological treatises,* vol. 1, pp. 12–22.

15  Some of the issues in this section have been dealt with before in Cunningham, 'Fabricius and the "Aristotle project"', 1985, from which I borrow freely.

16  Adelman, *Embryological treatises,* (1942), vol. 1, p. 25, shows that virtually every one of the anatomical topics on which Fabricius published was represented in his anatomical teaching for students, and of which they often complained.

17  *De Methodis*, cited by Randall, *The Career of Philosophy* (1962), p. 298.

18  Dedication to *De Formato Foetu*, Venice, 1600, reprinted in facsimile in Adelmann, *Embryological treatises* (1942) vol. 2, p. 464, my translation, based on that of Adelmann, ibid., vol. 1, p. 238.

19  On all these see my book *The Anatomical Renaissance.*

20  The following account is built up from what Fabricius says in different places in his writings, see especially *De Visu*, 1600.

21  These initial points, that is the *historia,* the action and the use of a part in anatomising, are built of course on Galen's anatomical writings. But Fabricius gives a distinctive Aristotelian reading of them, applying them not just to the body of man, but to 'The Animal'. Fabricius himself says that he has adopted this method of presentation 'the more willingly because those distinguished pioneers, Aristotle and Galen, have blazed the trail and, so to speak, carried the torch before me on my way'. Preface to *De voce*, as translated by Adelmann in *Embryological Treatises,* (1942) vol. 1, pp. 82–83.

22  On which see Jardine, 'Galileo's Road to Truth' (1976).

23  An alternative analysis of Fabricius' method and practice, but in the investigation of generation, has been offered by me in 'Fabricius and the "Aristotle project"' (1985), pp. 195–222.

24 My translation; for an alternative translation see Franklin, *De Venarum Ostiolis* (1933).
25 Dedication to *De Visu*, 1600.
26 Rippa Bonati, and Pardo-Tomás, *Il Teatro dei Corpi* (2004).
27 'Si quae vero admiratione digna Naturae habet contemplatio, ea profecto sunt quae ad animae facultates earumque instrumenta pertinent' (f. a.2). Speech, as such, is of course unique to humans, but he does also investigate the pertinent muscles in other animals.

# 4   William Harvey

## Pupil, physician, professor

### William Harvey, pupil: 1. England

William Harvey was born on 1 April 1578 in Folkstone in Kent, where his father was mayor on four occasions.[1] Harvey was educated at The King's School, Canterbury, and then he was awarded the Matthew Parker scholarship, which was restricted to boys from The King's School, and this took him to Gonville and Caius College in Cambridge to study philosophy and medicine. This college had been founded out of Gonville Hall, a somewhat decayed hall for students, by John Caius, a celebrated physician and the 'second founder' of the College of Physicians of London, in 1557. At that time Caius (as the college is known for short) was the most medical of all the colleges of Cambridge or Oxford, with more young men studying medicine there than at any other college, though still very few. Matthew Parker, Archbishop of Canterbury from 1559 to 1575, had been a friend of John Caius, and it is perhaps through this relationship that Parker had decided to found this medical scholarship – the first in England – at Caius College. William Harvey was the fourth student to hold it. He entered the college in May 1593 at the age of 15, staying until 1599, when he was 21. The chance of being sent to the right school, and being able to take this scholarship to Cambridge, thus determined Harvey's career as a doctor. It set his career path very firmly in medicine, unlike his brothers, five out of six of whom became Levant merchants.

This scholarship lasted for six years, and it would have been possible under the university statutes for Harvey to have spent the whole six years pursuing the bachelor's degree in medicine, without having to take an arts degree first. However, whilst Harvey did hold this scholarship for almost the full six years, he at some point decided to take the Bachelor of Arts degree in 1597. This suggests that he was primarily studying the conventional philosophy course, rather than medicine, at least for the first three years. He did not then proceed to take the bachelor's degree in medicine at the end of the sixth year of his scholarship. So it is not at all clear how much medicine Harvey studied at Cambridge – and hence whether Cambridge as his *alma mater* deserves any credit for training Harvey as the great anatomical researcher he turned out to be.

DOI: 10.4324/9781003247616-6

In Harvey's student days Cambridge was not a particularly good place to study medicine, though it was slowly improving. Cambridge was a university where the masters had for centuries been dominant in the organisation of teaching, and it was the regent masters or regent doctors, as they were called – those who had relatively recently graduated in each discipline – who were expected to do most of the teaching of the subject. This teaching consisted of the reading of authoritative texts out loud, commenting on them, and running disputations on them for the junior members of the faculty. All this was conducted in Latin. However assiduously it may be performed, such teaching leaves few traces. For medicine the texts being read out loud, commented and disputed on, would have been primarily those of two great ancient Greek physicians: some of the 30 or so anonymous works which go under the name of Hippocrates, and also some of the very extensive writings of Galen, which had been recently recovered, translated from Greek into Latin, and established as the basis of medical knowledge in the universities of Europe. Foremost among these works of Galen were his anatomical texts, and people had come to recognise in the course of the 16th century that Galen had had a formidable knowledge of human anatomy, which he had acquired from the dissection of animals, especially apes. The revival of this anatomical knowledge was perhaps the most important, and certainly the most spectacular, of the developments of medicine in the universities of Europe in the 16th century.

But Cambridge, like Oxford, was far from the centre of such activities. What we see here in England in Harvey's day are two relatively insignificant universities in a marginal country on the edge of European intellectual life, but where a few attempts were being made to bring them up to the standard of the best universities abroad, such as those of northern Italy and of France. One of the steps which had been made to make Cambridge more like an Italian or French university was the introduction of the professorial system and of paid lectureships. By Harvey's student days this meant that there was one professor of medicine at Cambridge, appointed by the crown, the 'Regius Professor of physic'. In addition to teaching Hippocrates and Galen, this professor was expected to perform anatomies, as would have been expected of such a professor on the continent, but nothing was done to secure bodies for him to dissect. A private benefactor, Thomas Linacre, a physician who had been at university in Padua and then had founded the College of Physicians in London, had endowed two lectureships in medicine, one at Cambridge and one at Oxford, and the Cambridge one was located in St John's College. This lecturer had to read Galen's books on the maintenance of health, on the method of healing, on the properties of foods, and on simples (that is, on plants used in medicines) 'to everyone wishing to hear'.

Such continental innovations had been further developed by John Caius in the Cambridge college he re-founded, and at which Harvey was a student. Caius himself had also studied medicine at Padua, in northern Italy, and whilst there he had participated in translating some of Galen's works from Greek to Latin, and had witnessed the re-establishing of anatomy as the basis of Galenic medicine. Caius certainly did what he could to promote the study of the latest – which was of course also the oldest, the Greek – form of medicine in his college. He not only provided for medical fellowships, and gave directions for the method of study to be engaged in by the medical fellows, but he also acquired from the crown the right to the bodies of two executed malefactors for the fellows of Caius College to dissect, twice a year, so that the medical fellows and students could learn the anatomy of the human. Twenty-six shillings and eight pence were to be spent every winter by those studying medicine in the college to carry out the dissection and bury the body afterwards with all due reverence. So, whether he saw anatomies performed there or not (which we do not know),[2] Harvey's own college actually had the best opportunities for witnessing human dissection of anywhere in England. In this sense Caius College deserves the compliment that Sir Charles Scarburgh gave it in a celebratory speech after Harvey's death. Scarburgh said that at Caius, Harvey 'drank in philosophy and medicine from the purest and richest spring of all, if there be such another dedicated to Apollo in the British Isles' (K. p. 14).

So we can say that at Caius College, Harvey was at the best possible place for learning academic medicine in the whole of England. However, his aspirations were considerably higher than this.

## William Harvey, pupil: 2. Italy

For in early 1600 Harvey arrived at the best university in the world for learning medicine, the University of Padua. As we saw in an earlier chapter, six years before, Shakespeare, through the mouth of Lucentio in *The Taming of the Shrew*, had called it 'fair Padua, nursery of arts' and, in its day, of all the arts Padua was the greatest nursery of medicine. It was here that William Harvey developed his passion for anatomical research, which was, all unknown to him, to lead him to discover the circulation of the blood. Just as he had been steered to Cambridge and Caius College by the chance of him first attending The King's School in Canterbury, so now, as a student at Caius College, he was assisted in his choice of Padua both by the example of John Caius himself, and also by an encouragement in the college statutes for the medical fellows to study abroad for three years at Padua, Bologna, Montpellier or Paris.

Shortly after arriving in Padua, Harvey was elected Consilarius for the English Nation, which meant that he acted as representative of part of the student body. His diploma for the degree of M.D. at Padua was awarded in April 1602 (K. pp. 32–33). He also visited the hospital at Padua, for a later note of his says 'These things I have examined both in the Hospital [of Saint Bartholomew in London] and in the Italian hospitals with much nausea and loathing and stench and I remember them, but many I have forgotten' (K. pp. 31–32).

This is all we know directly about his time in Padua, but as we shall see shortly his time at Padua was to set his whole career both in his medical practice and in his anatomical research. For Harvey attended the lectures and demonstrations of Professor Fabricius in anatomy and surgery, which we have discussed in Chapter 3. The only teacher-pupil reference to this that Harvey mentioned in print was when he wrote that he was now 'following the stepps of *Fabricius*, who was formerly my *Tutor*, and is now my guide' in exploring the generation of animals (*Generation*, p. 382).

## William Harvey, physician

Harvey returned to England in 1602 or 1603, and settled in London where he established a private medical practice. He married Elizabeth Browne, daughter of Dr Lancelot Browne, in 1604.

As a London medical practitioner, Harvey had to have the permission of the College of Physicians in order to practise; he was given a licence to practise in 1603, made a Licentiate of the College in 1604, and then joined as a full Fellow in 1607.

Harvey was a very committed member of the college, regularly attending the meetings at which the college proceeded against unlawful practitioners. He held all the offices of the college (Censor 1613 etc., Elect 1627, Treasurer 1628) except that of President, which he declined when approached in 1654.

In his later years, when he was famous throughout Europe, Harvey encouraged the other Fellows of the College of Physicians to undertake anatomical research together and to publish their findings, even leaving money in a Trust Deed for an annual feast and 'an Exhortacion to the ffellowes and members of the said Colledge to search and studdy out the secrets of Nature by way of Experiment' (K. p. 404). He donated a new building to the college as a library and depository for rarities and for the basic ingredients of medicines in 1651–1652, but this was burnt down in the Great Fire of London in 1666. It is clear that the College of Physicians played the role of a learned club for him, and that he held many of his colleagues there in high respect, for he was to consult them repeatedly about the disturbing findings he made about the movement of the blood.

Though young in practice, Harvey wanted an institutional position and with the help of his father-in-law tried to be appointed as Physician to the Tower of London in 1605. His father-in-law, Dr Lancelot Browne, spoke very highly of the medical knowledge of the young man:

> I dare assure you, that he is every way fitte for performance of as greate a charge as this is. I did never in my lyfe know any man any thinge near his yeares that was any way matche with him in all pointes of good learning, but especially in his profession of Physician wherein beinge examined in the Colledge [of Physicians] three several tymes, he did answere so readily and fully withall, as the whole company did both admire him, and took very singular lyking unto him.
>
> (K. p. 45)

But even with this glowing reference praising his medical knowledge young Harvey was not appointed to this position.

Some incidental or court records of Harvey practising as a physician in the early 1620s are recorded in Keynes' biography of Harvey.[3] John Aubrey recorded of Harvey's medical practice that

> All his Profession would allowe him to be an excellent Anatomist, but I never heard of any that admired his therapeutique way. I know severall practisers in this Towne London that would not have given 3[d] for one of his Bills and that a man could hardly tell by one of his Bills what he did aime at.
>
> (K. p. 435)

However, Keynes argues persuasively that Harvey's medical practice was simply not to Aubrey's taste (K. pp. 383–385). Moreover, Keynes shows that Harvey had many important patients, some of whom insisted on his attendance, whilst others thought his prescriptions too fanciful rather than practical. What direct evidence that there is indicates that Harvey diagnosed in a conventional Galenic way with humoral explanations, and resorted to the standard medical interventions of bleeding, purging and sweating his patients when indicated (K. pp. 294, 227). The editors of Harvey's Prelectiones manuscript have collected his incidental mentions of his medical and surgical practice as follows:

> Harvey claimed to have cured cases of diabetes, dropsy, venereal diseases, prolapsus uteri, varicose hernia, inguinal hernia, hydrocele, and many other conditions, and that he had even practiced surgery himself and had actually performed lithotomy.[4]

Harvey himself mentions performing post-mortems. He does so in the course of trying to persuade Professor Jean Riolan that blood must of necessity continually circulate or the heart would be oppressed by it, and death would follow. The first of these was on Sir Robert Darcie, a relative of Harvey's own best friend, Dr Argent. 'In his Corps, in the presence of Dr Argent, who at that time was President of the College of Physicians' Harvey discovered that the wall of the left ventricle of the heart was broken and 'poured forth blood at a wide hole', and that was the cause of death. Similarly he performed a post-mortem on 'a stout man, who did so boyl with rage … and was tortur'd, and miserably tormented with great pain and oppression in his heart and brest', and nothing the physicians could do would help. 'In his corps', Harvey relates, 'I found the heart and the aorta so distended and full of blood, that the bignesse of his heart, and the concavities of the ventricles, were equall in bignesse to that of an Oxe'. (*To Riolan*, pp. 62–64) We can I think assume that he performed other post-mortems as well. But Harvey did have two other long-term paying medical jobs in addition to his private medical practice. First, from 1609 he was physician at St Bartholomew's Hospital, for which he was paid £25 a year (later raised to £33), and for which he had to attend one day a week to care for the poor patients of the hospital. There were nurses in the hospital and three surgeons, but Harvey was the only physician. There were something around 200 beds in the hospital. This is how the hospital treasurer swore Harvey in, explaining his duties:

> You are here elected and admitted to be the Phisicion for the Poore of this hospitall, to perform the charge followinge, That is to say, one day in the weeke at the leaste thorough the yeare, or oftener as neede shall requyer you shall come to this Hospitall, and cause the Hospitler, Matron, or Porter, to call before you in the hall of this hospitall such and soe many of the poore harboured in this hospitall, as shall neede the counsel & advise of the phisicion ... wrytinge in a booke appoynted for that purpose, such medicines ... as apperteyneth to the apothecary of this house, to be provyded and made reddy for to be ministered unto the poore ... accordinge to his disease ...This you shall promise to doe as you shall answere before God, and as it becometh a faithful phisicion ...
>
> (K. pp. 53–54)

He retained this position for 35 years.

The second paid medical job Harvey held was as one of the royal physicians. He was initially appointed a physician extraordinary (the lower level) to king James I from before 1618, and was present at the death-bed of the king in 1625. James often ignored the advice of his physicians, laughing at medicine: his principal physician, Sir Theodore de Mayerne said that 'He asserts the art of medicine to be supported by mere conjectures and useless because uncertain' (K. p. 142). Harvey's appointment carried over to Charles I from his accession in 1625. From 1631 he was one of the Royal Physicians in Ordinary to the king, receiving an annuity of £50.[5] He was and remained a Royalist throughout the long period of the Puritan revolution (from the 1620s) and the civil war (1642–1649), and he even accompanied King Charles I on the battlefield in 1642.

These two jobs militated against each other somewhat. For as Charles called more and more on Harvey's services, especially when he went to Scotland in 1633 to be crowned, so Harvey perforce neglected his role at the hospital, and had to leave the work to his deputy, though he lost no income from the post (K. pp. 196–197, 202–203).

We need to recognise that as a full-time practising physician earning his living from his medical practice, Harvey was, necessarily, a follower of Galen, and this was not controversial but conventional. This is what he had been trained in at Padua. Whilst there were a few new approaches to medical understanding and practice being developed in his time, such as chemical medicine or spiritual medicine, Harvey had no time for them. As we shall see, Harvey adopted the Aristotelian programme in anatomical research. But there was no Aristotelian form of medical practice available for him to have chosen if he had wanted to, for Aristotle had not been a physician. So we have here in Harvey a man simultaneously in two ancient traditions and practices, one in his medical work and another in his anatomical research. In the one he was a Galenist, in the other an Aristotelian. It just wasn't possible for Harvey in his Aristotelian anatomical research to go back behind Galen entirely, so we should not be surprised to find Harvey using conventional Galenic language and concepts even whilst pursuing an anatomical research project which was radically different from that of Galen.

## William Harvey, professor

Given the distance of London from the great medical teaching centres of Italy and France, both in miles and in cultural sophistication, there was actually quite a significant amount of anatomical teaching supposedly available there offered by medical men in the early 17th century, at least intermittently. The College of Physicians nominally had an annual anatomical demonstration, and every Fellow was expected to be competent to deliver this (K. pp. 137, 311). Fines were often issued to Fellows by the college for failure to perform this (K. pp. 264, 277–278). The Company of Barber-Surgeons also provided anatomical teaching for their apprentices, provided by a physician; for instance from 1649 their reader was Dr Charles Scarburgh, who taught for 17 years (K. pp. 305, 307). By royal charters both the college and the Company were able to claim the bodies of hanged criminals for these demonstrations. But human bodies were not always essential for these; for instance the young Christopher Wren made pasteboard models for Dr Scarburgh to use (K. p. 305). In addition to these there was anatomical teaching to be had publicly at Gresham College; Dr Thomas Winston gave these for over 30 years from 1613 (K. p. 318), leading to the publication of his *Anatomy lectures at Gresham Colledge* in 1659, after his death.

A six-year round of lecturing and demonstrating on anatomy and surgery had been established in the College of Physicians by Lord Lumley and Dr Richard Caldwell and it began in 1584. This Lumleian lectureship was primarily intended to improve knowledge in surgery. The reader of this lecture was to be 'a doctor of physick, and of good practise and knowledge, and to have an honest stipend no lesse than those [Regius Professorships] of the universities erected by king Henrie the eight, namely of law divinitie, and physicke, and lands assured to the said college for the maintenance of the publicke lesson'. The lecturer or reader was to be paid £40 a year.[6]

This was the role that Harvey undertook for more than 40 years from 1615, his third paid job beyond his private medical practice. This appointment is what entitled him to call himself and be called 'Professor', though he was frequently not successful in obtaining the stipend.[7] On the title-page of his manuscript lectures first dated 1616 Harvey wrote in Latin that the lectures are 'by me, William Harvey, London physician, Professor of anatomy and surgery'. On the title-page of both of his anatomical books that were published in his life-time again Harvey stiles himself, or is stiled, *Professor*, the title for someone who teaches academically.[8] He is called 'Professor of Physick' in the 1653 English edition, and *Professor Anatomiae in Collegio Medicorum Londinensi* in the 1628 first publication of the book. Then in the *Anatomical Exercitations concerning the Generation of Living Creatures*, of 1653 he is called 'Doctor of Physick, and Professor of Anatomy, and Chirurgery, in the Colledge of Physitians of London. The title-page of the 1651 Latin edition says it is by 'Autore Guilielmo Harveo Anglo, in Collegio Medicorum Londinensium Anatomes & Chirurgiae Professore'.

He clearly had every right to call himself a Professor of Anatomy and Surgery, serving as he did in a professional college with increasing claims to expertise and authority. Interestingly, as we saw earlier, these two domains, anatomy and surgery, were the same ones in which Fabricius held his professorship for so many years at Padua University. If one were to subscribe to the centres/periphery interpretation of the spread of ideas in the past, then one could see Fabricius as the centre in Italy, being taken as a model by Harvey out on the periphery in England.

This further role that Harvey undertook for many years effectively meant that, on top of his other roles, he was a semi-professional anatomist, the only one in England. These lectures took place in the College of Physicians' dissection room, and were given initially to the apprentices of the surgeons, twice a week on a six-year cycle, with a five-day dissection of a specified region of the human body. French has claimed recently that the two anatomy lectures of the college took place in alternate years, and that there is no evidence that Harvey actually gave the surgical lectures, so that in practice he gave 'philosophical lectures for physicians'.[9]

This meant that Harvey had access to more human bodies to dissect than anyone else in Britain, and for a very long period of years. What is more, he was very assiduous in undertaking these lectures and dissections. And we know this because, astonishingly, the hand-written book of the notes which Harvey made for himself in order to give some of these lectures still survives. The notes in the 99 pages of the manuscript are in a mixture of Latin and English, with all the additions he made over the years. It seems to have been later acquired by Sir Hans Sloane, a great collector of the manuscripts of physicians, and is now in the Sloane collection at the British Library.

This manuscript was published in 1886 as an autotype of each hand-written page and with a transcription on the opposite page, and it was produced primarily at the cost of the Royal College of Physicians. There have been two attempts to translate this mixed Latin-English text into English, which were carried on simultaneously, but seemingly unknown to each other. The first was by a team of three Harvey scholars and appeared as 'an annotated translation' in 1961. Meanwhile Gweneth Whitteridge was finishing her version 'edited, with an introduction, translation and notes', which came out in 1964. She makes a single positive mention about the earlier translation.

The first of these translations says that the 1886 transcription 'deserves commendation' (p. 18) for the most part. The three Harvey scholars gained 'the impression that Harvey was not fully a master of the Latin language'

(p. 18) and often had to resort to English. By contrast Dr Whitteridge finds the 1886 transcription deplorable:

> Unfortunately, this transcript is full of errors on every page, and one is forced to conclude that the transcribers had sometimes little idea of the meaning of what Harvey had written. Words have often been misread, sometimes to the utter confounding of all sense ... there are also many mistakes in Latin grammar
>
> (p. xx)

on the part of the transcribers. And Harvey's Latin, she claims, 'is no better and no worse than that of many of his contemporaries' (p. xxi).

All this work from the transcriber of 1886 and the two translation projects was primarily aimed at finding the moment and date of Harvey first realising and then announcing that he had discovered the circulation of the blood. This is not a concern of mine, so we can leave aside here these issues of transcription and translation.

But what we have to note here is that the subject matter of these lectures was the *human* body and its treatment by surgeons. But meanwhile Harvey was also pursuing his own private anatomical research which was not focussed on the human but on *the animal*. It is certainly the case that anything Harvey discovered in his private research on the animal would, if relevant, be mentioned by him in his public lectures whose topic was the human.

## Notes

1   The exhaustive and highly reliable source for the surviving details of Harvey's life in both England and Italy is by Sir Geoffrey Keynes, *The Life of William Harvey*, Oxford, 1978, and like all other modern scholars of Harvey, I lean on it heavily here. References to it in the text of this chapter will be in the form of 'K. p. xx'.
2   Keynes, p. 11 says that years later Harvey referred to the small liver and spleen he had seen in a corpse at Cambridge.
3   Keynes pp. 113–125. Some of Harvey's prescriptions are on pages 437–445.
4   O'Malley et al., *Lectures*, p. 13.
5   On his payments for this role see Keynes, pp. 188–189, 279. Possibly he received £400 per year, see Keynes p. 280.
6   Cited from Holinshed's *Chronicles*, in C.D. O'Malley et al., *William Harvey Lectures on the Whole of Anatomy* (1961), pp. 3–5.
7   Keynes, p. 281; Harvey spent £500 on lawsuits fruitlessly pursuing this.
8   The *Oxford English Dictionary* sums up the academic usage of the title: 'A public teacher or instructor of the highest rank in a specific faculty or branch of learning'.
9   French, *William Harvey's Natural Philosophy* (1994), pp. 71–72.

# 5    William Harvey, searcher into the vegetative soul

> If you will enter with Heraclitus in Aristotle into a work-house (for so I will call it) for inspection of viler creatures, come hither, for the immortal gods are here likewise; and the great and Almighty Father is sometimes most conspicuous in the least and most inconsiderable creatures.
>
> (Harvey, *To Riolan 2*, p. 32)

Can we legitimately use the terms 'research' and 'researcher' for Harvey's work and motivation? 'Research' is more thorough than a search, it is re-search, the act of searching repeatedly and closely for something. The terms 'research' and 'researcher' were certainly in use in English (from the French) in Harvey's day. But one can certainly use the term 'searcher' for Harvey's projects of inquiry, and that is what I shall do here. Harvey himself used just this vocabulary in a Trust Deed to benefit the College of Physician in 1656, when he gave 'an Exhortacion to the ffellowes and members of the said Colledge to search and studdy out the secrets of Nature by way of Experiment' (Keynes, p. 404).

And then, how can one appropriately describe the nature of Harvey's investigative work on the animal body and Harvey's role in undertaking it? We cannot call him a scientist or a biologist because those terms and those occupations were not to be created until the 19th century. The large field of the enquiry into nature in Harvey's time was called 'natural philosophy' and was taught in universities throughout Europe. The philosopher sought for causes, and in natural philosophy (as opposed to, say, rational philosophy) the natural philosopher sought for causes of natural phenomena, as the name indicates. But he (it was always a he) did not have to engage in physical or practical investigation in his pursuit of causes: he could just as well use books as his sources, and build on these with his reasoning powers to reach the causes. Or he could trust in his own creative thinking. For instance Jean Fernel, celebrated French physician and writer on physiology, could in 1542 claim that the sensory information acquired from anatomy is merely the starting point for someone investigating how the human body functions:

> we shall start the beginning of the teaching of medicine from the human body, which is both the subject of the art of medicine and, first of all, it

DOI: 10.4324/9781003247616-7

comes most clearly under our senses. Then from there, led through all the minutiae, we shall be finally carried by an impulse of the mind, to those things which can be understood by thinking alone (*quae cogitatione sola comprehendi possunt*).[1]

René Decartes was arguing something similar in Harvey's lifetime, viz. that the senses are only the starting point for the investigator, and the more important stage of investigation is purely mental and philosophical. In seeking knowledge of nature, the head is still superior to the hand or the eye.

But Harvey was of a different persuasion. For whilst he would certainly have seen himself as a natural philosopher, and seen anatomy in general as part of natural philosophy, he was not just a theoretical philosopher but a practical one, indeed an *experimental* one. He believed that this approach gave better facts, and hence better explanations, and better understandings of causes. This will all be discussed further in Chapter 7 later.

What projects then do we know that Harvey was engaged in in his anatomical research? Thanks to the obsessional collecting bug of Sir Hans Sloane in the 17th century and to the investigative work of various Harvey scholars in the 20th century, we now have a number of surviving Harvey manuscripts identified, in addition that is to the three books published in his lifetime. This array now enables us to look at the range and extent of Harvey's anatomical research over his lifetime.

First there is the exception, that is, the work he had to do to prepare and to keep up-to-date his lectures at the College of Physicians. For this series was exclusively about the anatomy of the human body. This involved the dissection of humans, usually men and usually hanged for some crime. The manuscript of these lectures – Harvey's own manuscript from which he spoke and demonstrated – a text which is in both Latin and English, was identified in the British Library, and was printed in autotype (a photographic process) in 1886. More than one attempt has since been made to transcribe and translate these notes. Harvey gave these lectures probably every two years, adding notes to them over time, and the primary interest of these notes for historians of Harvey has been to try and identify the moment when he realised the blood must circulate. Here Harvey took as his guide Caspar Bauhin's *Theatrum Anatomicum*, which was published in 1605, with a second edition appearing in 1621.

As we did with his teacher Fabricius, let us lay out the research topics of Harvey that we know about. Some were published, some remained in manuscript and have been discovered since his death, and others he himself refers to in his published works but no actual manuscripts of them appear to have survived. Let us lay them out, as we did with Fabricius, in accordance with the soul anatomy of Aristotle (see Chapter 3).

Obviously there was one major project on the movement of the heart and blood, and another on the generation of animals, because these resulted in two of his three published books.

| Aristotelian aspect of soul | Works by Harvey |
|---|---|
| Vegetative Soul | *On the Movement of the Heart and Blood in Animals*<br>*On the Generation of Animals*, including<br>*On insects*<br>I remember I have heard him say he wrote a booke de insectis, which he had been many yeares about, & had made curious researches and anatomicall observations on them; this booke was lost when his lodgings at Whitehall were plundered in the time of the Rebellion: he could never for love nor money retrieve them or heare what became of them and sayed *'twas the greatest crucifying to him that ever he had in all his life*. Aubrey (K. p. 436). See also *Generation*, p. 418.<br>*On Love, Lust*, and *act of Generation of Animals*<br>But of this [how male birds know females are ready for coition] elsewhere, in our tract of the *Love, Lust*, and *act of Generation of Animals*, we shall treat at large. *Generation*, p. 34.<br>*On respiration*<br>In *On the generation of Animals*, Ex., 3, pages 6 and 8, he mentions his '*Exercitations of the causes, instruments, and uses of respiration*'. One of Harvey's discoveries in this (lost?) work is '*The Perforation* of the *Lungs*, by me first discovered (of which I made mention but now) ... in *Birds* especially, very patent ... In ... almost all *Birds* whatsoever, a probe being put into the *winde-pipe*, will finde an open and wide passage, clean through the *lungs* ...<br>But to determine and give a reason of this is nothing else but a search for what the *lungs* were made ... I shall leave these things fitter to be set forth in a Treatise by themselves'<br>*Motion*, Ch. 6, pp. 34–35 |
| Sensitive Soul<br>Motile Soul | *Local motion of animals* and *Muscles*<br>'these things [contractions] are more fully and openly to be declared elsewhere, in the organs of motion of living creatures'. *To Riolan 2*, p. 53. See also *Motion*, Ch. 17, p. 103. Manuscript edited and published by Whitteridge. |
| Rational Soul | Conceptions of the brain and womb<br>'But how this Immateriall cause, as the principle, can be alike in the Braine, and in the *Uterus* ... and what the difference betweene them is: hereafter, **when we shall treat Universally of the Generation of all** *Animals* (even of those also, which are generated by *Metamorphosis*; namely, of *Insects*, and *Spontaneous Productions* ... and also **when we shall discourse of the Soule, and its affections;** and also how Arts, Memory, and Experience, are onely the Conceptions of the Brain, we shall endeavour both largely, and perspicuously to explaine'. *Generation*, pp. 565–6. See also ibid., p. 270. |

| *Aristotelian aspect of soul* | *Works by Harvey* |
| --- | --- |
| Other ongoing projects | *Medicinal Anatomy*<br>[encouraged by Riolan's example] Harvey intends 'to put forth and joyn my medicinal Anatomie ... that I might undertake to relate from the many dissections of sick persons, in what manner, and how the inward parts of them are chang'd, in place, bignesse, condition, figure, substance, and other sensible accidents, from their natural form and appearance ...' *To Riolan 1*, p. 2.<br>*Discourses of Pathology* and *Physiology*<br>'I may set down such things in my medicinal observations, and discourses of Pathology'. And a Treatise of Physiology: '... Spirits, which we ought to define, and show what and how they are in a Treatise of Physiologie, only I will adjoyn.' *To Riolan 2*, pp. 85 and 48. |

All this shows two significant things about the nature of his anatomical research. One is that the range of topics is very similar to that of his teacher Fabricius. Second is that almost all of it was – like the research of Fabricius himself – Aristotelian in concept and approach. This in turn makes it possible for us to put aside the old supposed 'problem' about his work, that there two William Harveys, an earlier one who wrote the book on the heart and was modern, and a later one who wrote the book on generation and was reactionary.

As can be seen from this comparison, most of the research areas that Harvey pursued in anatomy concerned the 'vegetative soul', or vegetable soul or vegetal soul (I shall use these terms interchangeably).

What was the vegetable soul? We keep meeting the vegetable soul in the course of this book, especially in the case of Fabricius. The concept dates from Aristotle, and was still current in Harvey's day, and far from being unique to Harvey. It's something that almost defies definition in modern terms. In the 19th century George Henry Lewes, experimental physiologist and historian, described it like this in his book on Aristotle of 1864:

> the Vegetal Soul, which is supposed to be the cause of all the phenomena observed in plants, is not a plant, nor a property of the plant, nor the resultant of the plant's many properties; it is an existence *sui generis*, in virtue of which the plant *is*. At the same time this Vegetal Soul is exclusively limited to the plant; it has no other form of existence; it exists only under the conditions of plant-life ...
>
> (p. 30)

The vegetal soul not only *is* what makes the plant the plant, but it also controls its plant-ness and its development and reproduction as a plant. In particular it is the source and centre of its nourishment and its growth. As Lewes puts it, it is supposed to be 'the cause of all the phenomena observed in plants'. This very nicely indicates why it is a philosophical issue, for in the Greek tradition the central role of the philosopher was to seek the causes of phenomena.

The vegetal soul is the most fundamental of the life souls, so animals (including humans) also have a vegetal soul controlling their digestion, distribution of food, nutrition, expulsion of wastes, breathing, heartbeat and the whole course of the generation of one animal from another. Basically the vegetal soul controls all the continuing functions of the insides of an animal keeping it alive and functioning. For humans, from the first breath we draw in as newborn babies, to the last breath we breathe out on our deathbeds, from the moment we swallow food or drink until it is expelled through the usual channels, from our first stirring of lust through to mating and gestation to the birth of another human being, everything is controlled by the vegetable soul. And the centre of all these actions, and of the growth and development of animals was, for followers of Aristotle such as William Harvey, the heart.

Today we use the expression 'vegetative' to describe someone who is still alive, after a major accident for instance, but who shows no brain activity. They continue to live but only as a plant, as it were.

Again, today we might use the term 'the autonomic nervous system' to cover the wide range of inner activities undertaken and steered by the Aristotelian vegetable soul. But, as our modern term shows, our understanding of these activities is through the nervous system, and thus quite different from the workings of the vegetal soul.

What happened to the vegetable soul after the mid-17th century? Lucas John Mix has recently put it very well:

> In the Renaissance and Enlightenment, theories of life changed dramatically. Cartesian dualism pushed the human soul out of the natural world, along with will, reason, and agency. Vegetable and animal souls, meanwhile, disappeared. They were neglected in the humanities and intentionally ejected from the natural sciences. Empiricists considered them unknowable, along with formal and final causes. The move proved tremendously successful in the development of modern biology, but it prohibited any common understanding that would bridge human and non-human life.[2]

With respect to William Harvey and his anatomical research we know that his thinking is shaped by the concept of the vegetable soul. In the first place, as we saw above, his research topics match those of Fabricius and Aristotle. But in the second place in his published works he himself talks repeatedly about the vegetable soul.

The first of these is his Dedication of his book on the motion of the heart and blood in animals. In the very first sentence he uses the concept to address his patron the King, Charles I, as the heart of his kingdom:

> Most Gracious King, The Heart of creatures is the foundation of life, the Prince of all, the Sun of their Microcosm, on which all **vegetation** does depend, from whence all vigor and strength does flow. Likewise the King

is the foundation of his Kingdoms, and the Sun of his Microcosm, the Heart of his Common-wealth, from whence all power and mercy proceeds.

And again, writing to Professor Riolan, and talking about the heating role of blood circulation:

> This indeed is the chief use & end of the Circulation of the blood, for which cause, the blood by its continual course, and perpetual influence, is driven about; namely, that all the parts depending upon it by their first innate warm moisture might be retain'd in life, and in **their own vital and vegetative essence**, and perform all their functions, whilst (as the Naturalists say) they are sustain'd and actuated by natural heat, and vital spirits …
>
> (*To Riolan 1*, p. 17)

The next and most important occasion is in Harvey's Preface to his book on the generation of living creatures, where he writes:

> We therefore (according to the Method proposed) will explaine, first in an *Egge*, and afterwards in other *Conceptions* of several creatures, what is constituted *first*, and what *last*, in a most miraculous order, and with a most inimitable prudence and wisdom, by the great God of nature; and at length we will discover, what we have found out, concerning the *first matter* out of which, and the *first efficient* by which, the *foetus* is made, as also of the *order & Oeconomy* of *Generation*: that thence we may attain to some infallible knowledge of each faculty of **the formative and vegetative Soul**, by the effects of it; and of the nature of the *Soul* it selfe, by the *parts*, or *organs of the body*, and their functions.
>
> (Preface, b 5–6)

And again,

> Having thorough insight & knowledge of these things, we may then contemplate the abstruse nature of the *Vegetative Soul*; and discern in all creatures what ever, the *manner, order* and *causes of their Generation* …

How does Harvey go about his enquiry into the generation of animals, and thus explore 'the formative and vegetative soul'?[3]

Before we start it needs to be said that Harvey's book on the generation of animals is conceived as a single over-arching narrative, it is very well written, engaging, well-paced, occasionally amusing, and personal in the way Harvey describes his feelings as he investigates and his hesitations, his deliberations, and the occasions when he changes his mind, writing for instance: 'And here my thoughts were a long time strangely divided, what I should resolve concerning the *Candidum colliquamentum*' (p. 86), or 'Having often observed these things with much caution and circumspection in several *eggs*, I stood awhile in suspense, what opinion I should entertain of them' (p. 107), or again: 'I can scarce refrain my pen from rebuking those that follow *Empedocles* and *Hippocrates* also' (p. 467).

The whole time Harvey is working on his own treatise on the generation of animals, he has constantly before his eyes the texts of Aristotle *On the Generation of Animals* and his *History of Animals*, especially Books 6 and 9, both in Latin, and for much of the time he also has the Latin text by Fabricius *De formation ovi et pulli (On the formation of the egg and hen)*, published posthumously at Padua in 1621. He compares what they each have to say, passage by passage, with every step of his own observational and experimental work, and quotes them extensively. This practice mirrors that of Fabricius, who had the texts of Aristotle's books before *his* eyes when he performed his investigations back in Padua. Both investigations (Fabricius' and Harvey's) take the domestic chicken as their model animal, opening the fertilised eggs over the course of days to follow the development of the chick. The chick is becoming alive in the egg, and Harvey is seeking to establish the precise moment that this happens. Harvey also has the writings of others to hand. In his *Tables of the Principal External and Internal Parts of the Human Body* (Nuremberg,1572) Volcher Coiter recalls that in May in 1564 at Bologna, with his teacher Ulysees Aldrovandus, he had opened the fertilised eggs of two clucking hens on successive days, primarily to discover what is the origin of the veins and what part is first formed. Aldrovandus investigated birds, producing his *Ornithology* in 1600 at Bologna, and discusses and illustrates the development of the chick in the second volume of this three-volume work. According to the historian Adelmann, both Aldrovandus and Coiter mention 'the question of the primacy of the parts in development as a motive for undertaking their investigations'.[4]

In his finished treatise Harvey explores everything as a dispute between, on the one hand, Physicians, that is, medically-trained followers of Galen, then and now, and, on the other hand, Philosophers, that is, followers of Aristotle, then and now. Both had written plenty about the issues, Galen primarily about humans, Aristotle about all animals. Both Aristotle and Galen had conducted experiments, but Harvey starts by announcing that what both schools thought about it was all wrong, and that he had come to his conclusions 'out of Anatomical dissections' (Preface, p.1). He urges the Reader to trust their eyes, 'when the light of Anatomical dissection breaks forth'.

According to Harvey the Galenists think that the male seed is the *efficient* cause, the female seed is the *material* cause; and they also think 'the clean contrary'. Aristotle says the female contributes the *matter*, the male the *form*, and immediately after coition 'is formed in the Womb out of the Menstruous bloud, the Vital principle, and first particle of the future *Foetus*, (namely the Heart, in Creatures that have bloud)'. If I may leap to Harvey's conclusion about this complicated quarrel, it is going to be that *both* the male and female *equally* contribute form, species and soul (Ex. 29, p. 162). He can conclude: 'Now, having made strict discovery, by *Anatomical dissection*, of the parts sacred to *Generation*, we well know, what both *Male* and *Female* doe conferre to it. For the knowledge of the *Instruments* doth lead in a straight line, to their *functions* and *uses*' (Ex. 38, pp.188–9).

More generally Harvey begins: 'in the Generation of Animals (as in all other things of which we covet to know anything) every inquisition is to be derived from its *Causes*, and chiefly from the *Material* and *Efficient* ...' (p. 17). 'Derived from its causes' means 'traced back to its causes', or (as the Latin translation has it) 'sought from its causes' (*inquisitio omnis a caussis petenda est*).

What Harvey is looking for, as he repeatedly says, is (1) 'what is constituted *first*, and what *last*' in the generation of animals; (2) the *first matter* out of which, and the *first efficient* by which, the *foetus* is made'; and (3) 'the *order & Oeconomy* of *Generation*'. Throughout his work the *material* and the *efficient* causes (two of Aristotle's four causes of explanation of natural phenomena) are of the greatest concern to Harvey.

Right at the beginning Harvey announces one of his most important conclusions:

> we pronounce all animals whatever, even *Viviparous* also, nay *man* himself to be made of an *Egge* ... The history therefore of *Egges* is most spacious, because it yields an insight into all kinde of *generation*. Wherefore of an *Egge*, we shall first shew, *where*, *whence*, and *how* it is made. And then, by what *means*, *order*, and *degrees*, the *foetus* or *chicken* is fashioned, and perfected in the *Egge*, and of it.
>
> (p. 2)

Beginning with the courtship and mating of animals, and especially of course with birds and hens as they are Harvey's model animals, he explores the anatomy of the hen and how it receives the seed from the male (if indeed seed is emitted and/or is conveyed across to the female). Then he traces the stages of the formation of the egg in the hen, seeing how the yolk falls from the cluster, through the infundibulum and '(by virtue of a *vegitative* heat, and faculty wherewith it is endowed) findeth out, and concocteth its *white*' (p 44).

Once the egg is laid containing – as can be seen by breaking it – the yolk, the white and the membranes, inside it the chick begins to develop: and apart from the warmth of the hen sitting on the egg, all these processes are carried out by the operations of the vegetative soul within the egg itself.

The moment, the most crucial moment, that Harvey is particularly interested in is the moment when 'the *Egg* beginneth to step from the life of a *Plant*, to the life of an *Animal*' (Ex. 17, p. 89), and this centres of course on the first appearance of the blood, of pulsation in that blood, and then the appearance of the blood vessels and the heart. This begins to be visible on the third inspection of the egg, that is three days and nights after it was laid: 'You will meet a great *Metamorphosis*, and wonderfull alteration', Harvey writes to his reader (Ex. 17, p. 89). What Harvey has seen 'and so have many more who have been present' (Ex. 17, p. 95) confirms what Aristotle says about this: 'that the *Egge* which was before endowed with a *vegetative soule*, is now over and above that, furnished with a *Motive* and *sensitive power*; and is raised from a *Plant* to an *Animal*' (Ex. 17, p. 94).

This is what Aristotle wrote about the *punctum saliens*, the capering bloody point, in *History of Animals,* book 6. c. 3:

> During this period the yolk has worked its way up towards the pointed end, where the 'principle' of the egg is situated and where the egg hatches; and the heart is no bigger than just a small blood-spot in the white. This spot beats and moves as though it were alive; and from it, as it grows, two vein-like passages with blood in them lead on a twisted course to each of the two surrounding envelopes.

Now what Aristotle says here is 'the heart is visible like a red spot', and that it palpitates and moves. Fabricius had seen the palpitating red spot in his investigations, and illustrated it in one of his tables, says Harvey, but '(which is strange) took it for the body of the *Foetus*' (Ex. 17, p. 90). Aldrovandus, writing on birds in 1600 reported having seen it, as did Coiter, similarly writing on birds in 1573, but neither had interpreted it correctly, in Harvey's view.

On the fourth day this leaping bloody point is more visible. Harvey writes:

> For now the *Limbus* or *hemme* of the *colliquamentum* beginneth to blush and purple, being encompassed with a slender bloody line: and in the center almost of it, there leapeth a *capering bloody point*, which is yet so exceeding small, that in its *Diastole*, or Dilatation, it flasheth only like the most obscure and almost indiscernible spark of fire; and presently upon its *Systole* or Contraction, it is too subtile for the eye and quite disappeareth.
>
> (Ex. 17, p. 89)

At the end of the fourth day it is

> most notoriously visible, *that the* Punctum sanguineum saliens *hath now Animal motion (*saith *Aristotle) in the candid liquor* (which I call *colliquamentum*). ... In ... the Middle of the *colliquamentum,* the *Punctum Rubrum Saliens* is enthroned, which keeps time and *decorum* in its *pulsation.*
>
> (Ex. 17, p. 91)

For Harvey, 'This disquisition is of great moment; namely, Whether there be *blood* before *Pulsation*, and Whether the *Punctus* arise from the *Veines,* or the *Veines* from the *Punctus?*' (Ex. 17, p. 92).

We can say that by the end of the book Harvey has done what he set out to do, that is, go back to the beginning of generation and

> explain ... what is constituted *first*, and what *last* ... and at length we will discover, what we have found out, concerning the *first matter* out of which, and the *first efficient* by which, the *foetus* is made, as also of the

order & *Oeconomy* of *Generation*: that thence we may attain to some infallible knowledge of each faculty of **the formative and vegetative *Soul***, by the effects of it; and of the nature of the *Soul* it selfe, by the *parts*, or *organs of the body*, and their functions.

(Preface, b 5–6)

No-one before had done this in such detail and with such precision of observation and experiment.

Harvey's conclusions here are quite momentous. By tracing, with daily experiments and viewings, the whole process of the development of the chick in the egg, he can say:

It therefore is clear by our *History*, that the *generation* of the *Chicken* out of the *Egge*, proceeds rather *per Epigenesin, quam per Metamorphosin*, by an *Epigenesis*, then by a *Metamorphosis*; and that all its parts are not constituted at once, but successively, & in *Order*; and that while it is augmented it is also formed, & while it is formed, it is also augmented ... *Epigenesis*, in which an order is observed according to the dignity, and worth, and use of the *Parts*.

(Ex. 45, pp. 224–226)

What then appears first, what is the order of generation? Harvey argues that the blood is the first genital particle:

And hence the prerogative and antiquity of the *blood* appears, seeing that the *Pulse* proceedeth from it (Ex. 51, p. 275) ... '*Life* therefore consists in the *blood*, (as we read in *Holy Scripture* [Levit. 17, 11 & 14] because in it the *Life* and *Soule* do first dawn, and last set.

And frequent dissection of live animals shows that it is the right auricle of the heart which is the last part to die (Ex. 51, p. 277; Ex. 18, p. 108).

Harvey's oratory here shows just how important the blood is in the generation of animals.

By all which it is most evident, that the blood is the *Genital Part*, the fountain of Life, *Primum vivens, & ultimum moriens*, the First-born, and the Longest Liver, and the chief Palace and Court of the *soul*: in which (as in its Spring-head) the heat doth first and chiefly flow, and flourish: and from which all the other *parts* of the *Body* derive their life and influent warmth. For that heat streaming with the blood, doth sprinkle, cherish, and preserve the whole: as we have heretofore demonstrated in our Booke, *de motu sanguinis*.

(Ex. 51, p. 278)

From Exercise 64 Harvey, having now dealt with the egg and the chick, turns to animals which give birth to perfect live progeny: viviparous animals. As he took the chicken as his model animal for egg-laying animals, he now takes hinds and does 'as the Example of all other Animals', as the chapter-title has it. His reason for this, as he explains, is that, unlike chickens, it is difficult to have repeated access to large animals to slaughter at different intervals after coition in order to inspect the progress of the development of the offspring. However, as a royal physician to Charles I, Harvey has privileged access to the royal hinds and does in Richmond Park.

> Hereupon (for the *Rutting time*, when the *Females* are lusty, and admit the *Males*, whereby they conceive and bear their young) I had a daily opportunity of dissecting them, and making inspection and observation of all their *parts*; which liberty I chiefly made use of in order to the *Genital parts*.
>
> (Ex. 64, p. 397)

In all this he was encouraged by his Royal Master

> who was himself much delighted in this kind of curiosity, being many times pleased to be an eye witness, and to assert my new inventions). I had great store of his *Deere* at my devotion, and frequent opportunity and license to dissect and search into them.
>
> (Ex. 64, p. 397)

Of viviparous animals, man

> is the most consummate or complete *Animal* of all other, as he hath obtained all other *parts* more perfect then they, so are his *Genital parts* also. And therefore the *Uterine parts are* most distinct in a *Woman*, and to us (by reason of the special industry of *Anatomists* about this *Part*) better known
>
> (Ex. 65, pp. 398–399)

So in a way Harvey takes human women as his alternate study. Whilst he cannot, of course, dissect them after coition he can nevertheless dissect miscarriages, spontaneous abortions of different dates, and can observe much from his practice with women at and after childbirth, and dying before giving birth. He compares the parts of the womb of women with his does, 'by their situation, connexion, largeness, perforation, form and function', things which 'I have seen with my own eyes' (Ex. 65, p. 403; Ex. 56, pp. 334–9).

The finding which most fascinated the King and most disturbed the keepers, huntsmen, and Harvey's fellow physicians, was that in the womb

> I could never discover any *Sperme* ejected from the *Male Deere*, nor any other thing which relates to the *conception* … And yet the *Male Deere* did go to rut daily, and I dissected a great number of *Does*. And this is the result of many years experience
>
> (Ex. 67, pp. 412–413)

The keepers and huntsmen at last 'being confuted by their own eyes, they sate down in a gaze and gave it over for granted'. But the physicians could not be persuaded (Ex. 68, pp. 416–7).

The *punctum saliens*, the 'capering point' of the first visible blood, Harvey can detect in the conception of the doe as he had in the chick in the egg.

> Having dissected the *uterus* [of a doe], I have exposed this *Punctum saliens*, while it yet continued its palpitation, to the view of our late dread *Soveraigne*; which was then so small, that without the advantage of the Sun-beams obliquely illustrating it, he could not have perceived its shivering motion.
>
> (Ex. 69, p. 423)

When Harvey discusses the innate heat, his text is nothing less than a paeon of praise to the blood, its many roles and faculties, unlike the elusive spirits if they even exist.

> The *blood* therefore, as well as the *soul*, is to be reputed the cause and author, both of Youth, and Old Age, of Sleep and Waking, and of Breathing also ... And therefore it comes all to the same reckoning, whether we say, that the *soul* and the *blood*, or the *blood* with the *soul*, or the *soul* with the *blood*, doth performe all the effects in an *Animal*.
>
> (Ex. 71, p. 460)

Every step Harvey makes in his investigation of the generation of animals brings his attention back to the overwhelming importance of the blood. It appears first, it physically constructs its containers, the heart and the blood vessels, it constructs all the other parts of the body – even the liver, which Galen and the physicians had thought made the blood! 'Without all question the *Blood* is to be counted the *Author* of the *Liver,* rather then the *Liver* the *Author* of the *Blood.*' (Ex. 19, p. 115; Ex. 46, p. 232)

It is the soul, both the vegetative and the animal soul. As will appear in Chapter 8 below, the blood as such had not hitherto been an object or topic of research for anatomists, nor even much of a topic for the theoretical musings of physiologists. But now, with Fabricius to a certain extent, but with Harvey above all, the blood is life and the soul, and its container, the heart, is confirmed as being very clearly the centre of the vegetative soul. He writes:

> Being therefore ascertained out of those things which I have observed in an *Egge*, and the *dissection* of *Animals* while they were alive, I conclude (against *Aristotle*) that the *blood* is the first *Genital particle*, and that the *Heart* is its Instrument designed for its *Circulation*. For the *Hearts* business or function is the propulsion or driving forth of the *blood*, as appears in all *Animals* that have *blood*: and the office of the *Vesicula pulsans* is the very same, (in the *generation* of the *Chicken*) which I have shewed to many persons, in the first *conceptions* of *Animals*, (as well as in an *Egge*)

when it hath been less then a *Spark*, panting, and in its motion, drawing it self together, and so squeezing out the *blood* contained in it, and by relaxing itself again, receiving and entertaining *blood* afresh.

(Ex. 51, p. 275)

He has repeatedly made this experiment of the *vesicula pulsans* in the egg and in the conceptions of viviparous animals, and he has shared what he could see with many other people. This is Harvey the experimenter at work, and shows him relying on his senses, not on authority – even the authority of Aristotle. And this great statement by Harvey on the role of the heart in all animals that have blood, and of the prerogative and antiquity of the blood with respect to the heart, leads us on naturally to discussion of Harvey's first book and the project on which it reports, about the motion of the heart and blood in animals – another book on the vegetable soul.

## Notes

1 Fernel, *Medicina*, p. lv, my translation.
2 Mix, *Life Concepts* (2018), p. 2. See also Baldassarri and Blank, *Vegetative Powers* (2021).
3 A good summary of the book is given by Meyer, *An Analysis*, and a valuable discussion of the structure of the argument is presented by Whitteridge in the Introduction to her translation. The best account of what Harvey is doing and finding (and not finding) in his researches is still that given by Gasking, *Investigations into Generation* (1967), chapter 1.
4 Adelmann, *Embryological Treatises*, vol. 1, pp. 67–9; the section in Coiter's book is pp. 32–39.

# 6    'The wonderful circulation of the blood first found out by me'[1]

> all difficulties cease, when there are not two contrary motions supposed
> in the same vessels; but that we do suppose that there is one continued
> motion.[2]

So we come at last to the heart of the whole issue of how William Harvey
discovered the circulation of the blood. We have put aside the customary
view that the discovery was somehow a race, even a very slow race over centuries. Similarly we have put aside the view that it was some sort of puzzle, in
which over the centuries various thinkers had put forward one or other piece
of the puzzle and our Harvey finished the puzzle with the final few pieces.
Again, we have put aside the view that anyone – anyone at all – had been
looking for evidence for the circulation of the blood. So we don't need to
hunt for any predecessors in the business because there weren't any. We now
just need to keep our eyes on Harvey, on what he was doing and how, and on
what he said in his book and elsewhere announcing the unexpected and most
unwelcome discovery.

In this context, we need to ask what strange enquiry Harvey must have
been engaged on that led him to see things completely differently from his
contemporary physicians, the Galenists. It was not, as some older historians
used to maintain, simply a matter of Harvey using his eyes better than other
researchers, nor the fact that he used experiment, for his opponents and rivals
did this too. Indeed, with respect to using one's eyes, the circulation of the
blood is not something which is visible: it is the *deduction* of an argument
about anatomical pathways, the rate of pulsation, the capacity of the chambers of the heart, the competence of valves and their location, and so on. The
circulation of the blood is certainly *not* something lying in front of one's eyes,
just waiting to be found!

On the positive side we now know securely that Harvey was an Aristotelian,
and that in this he followed his own anatomical teacher in Padua, Fabricius.
We also know from his practice that as an active researcher he very much wants
to be part of this tradition, even though he has very little immediate support,
given that in England he was living on the fringes of the European medical
and anatomical world. So he would obviously choose anatomical topics to

DOI: 10.4324/9781003247616-8

investigate which flowed from this Aristotelian/Fabrician investigative programme, as we have seen with respect to his work on the vegetative soul and the generation of animals.

Whatever we might have thought earlier about how 'modern' Harvey is, his work on the movement of the heart and blood in animals actually *is* on the vegetative soul! Recall the first sentence of the Dedication of his book to his patron the King, Charles I, as the heart of his kingdom:

> Most Gracious King, The Heart of creatures is the foundation of life, the Prince of all, the Sun of their Microcosm, on which all **vegetation** does depend, from whence all vigor and strength does flow. Likewise the King is the foundation of his Kingdoms, and the Sun of his Microcosm, the Heart of his Common-wealth, from whence all power and mercy proceeds.
>
> (My emphasis)

In the previous chapter we saw other usages of this term by Harvey, in his letters to Professor Riolan and in his book on the generation of animals. But here in the book on the motion of the heart and blood in animals Harvey again refers to the actions of the vegetative soul, first in a sentiment confirming his Dedication:

> So the *heart* is the beginning of life, the *Sun* of the *Microcosm*, as proportionably the *Sun* deserves to be call'd the *heart* of the world, by whose virtue and pulsation, the blood is mov'd, perfected, made **vegetable**, and is defended from corruption, and mattering; and this familiar household-god doth his duty to the whole body, by nourishing, cherishing, and **vegetating**, being the foundation of life, and author of all.
>
> (My emphasis; Ch. 8, p. 47)

Also, for an animal to remain alive, Harvey argues,

> there must needs be a place and beginning of heat ... from whence heat and life may flow, as from their beginnings, into all parts; whither the aliment of it should come, and on which all *nutrition* and *vegetation* should depend. And that this place is the *heart*, from whence is the beginning of life, I would have no body to doubt.
>
> (My emphasis; Ch. 15, pp. 81–82)

And vegetivity, the work of the vegetable soul, was indeed for Harvey – finally – the *cause* why the blood circulated. Writing to Professor Riolan in Paris in 1649 he concluded that

> This indeed is the chief use & end of the Circulation of the blood, for which cause, the blood by its continual course, and perpetual influence, is

driven about; namely, that all the parts depending upon it by their first innate warm moisture might be retain'd in life, and in their own vital and **vegetative** essence, and perform all their functions, whilst (as the Naturalists say) they are sustain'd and actuated by natural heat, and vital spirits ...

(My emphasis; *To Riolan 1,* p. 17)

'I follow Aristotle', Harvey proudly proclaimed. My whole thesis here in this book is that (1) Harvey discovered the circulation of the blood because he was an Aristotelian anatomist, and (2) that only an Aristotelian anatomist could have done so. If my thesis is to stick, what criteria would have to be satisfied for me to be able to claim that Harvey's work in this area was Aristotelian, like Aristotle himself and like his own teacher Fabricius? These criteria would also have to be ones which distinguished this Aristotelian tradition of anatomising from other traditions of the time.

1.   In the first place, following my exposition of Aristotle and animals, the project would have to be about the *soul*.
2.   Second, it would have to be a project seeking *causes*, causes why particular anatomical structures are there and why they do what they do – their functions. Another way of putting this would be: a project investigating certain functions of the animal and the organs which carry them out. As Aristotle wrote in *Parts of Animals*, 642 a–b: 'Of the method itself the following is an example. In dealing with respiration we must show that it takes place for such or such an object [= final cause]; and we must also show that this and that part of the process is necessitated by this or that other stage of it.'[3]
3.   Third, it would have to be a project which dealt not with one species or kind of animal, but with all animals, or 'the Animal' – and not just with, for instance, man.
4.   Fourth, the project would be fulfilled when variations of the main answer in different animals can be explained by specifying one or more of four aspects of the creature in question: its life, its activities, its habits or its (other) parts.

I hope that my reader will feel satisfied from my account that this is what Aristotle himself does – as well as advocates – and what Fabricius also does in his *Theatre of the Whole Animal Fabric*. It is also what Harvey did in his work resulting in his book *Anatomical Exercitations Concerning the Generation of Living Creatures*, as we saw in the preceding chapter. The crucial question to be satisfied now is whether Harvey did the same in his work resulting in his book the *Anatomical Exercises Concerning the Motion of the Heart, and Blood, in Living Creatures*.

Harvey's project, the one which led him unexpectedly to discover the circulation of the blood, appears in three formulations in the book, and where one might expect to find them, that is at the beginning:

(1) seeing 'we are thinking of the motion, pulse, use, action, and utility of the *heart* and *arteries'*. [Proeme, opening sentence]
(2) 'it will be profitable to search more deeply into the businesse, and to contemplate the motions of the *arteries* and *heart*, not only in man, but also in all other creatures that have a *heart*; as likewise by the frequent dissection of living things, and by much ocular testimony to discern and search the truth'. [Proeme, final sentence]
(3) 'When first I applyed my mind, to observation, from the many dissections of Living Creatures as they came to hand, that by that meanes I might find out the use of the motion of the *Heart* and things conducible in Creatures; I straightwayes found it a thing hard to be attained, and full of difficultie'. [Chapter 1, opening sentence]

The project from which his discovery of the circulation of the blood resulted was thus one drawn up on strictly Aristotelian/Fabrician lines. For as we see, it was about 'the motion, pulse, use, action and utility of the *heart* and *arteries'* in living creatures – the heart being the central organ of the vegetative soul.

So that is where the project started: with the heart and its pulsation, and with the arteries and their pulsation. Whitteridge thinks that 'by 1616 [when he started giving his lectures] Harvey's observations on the movement of the heart were practically complete' (*Praelectiones*, xliii). And yet, the book that he published 12 years later in 1628 had a title announcing something else: the motion of the heart and *blood* in animals. To this radical change of topic we shall return.

But why all this attention on the heart by Harvey? Roger French writes:

> We have no reason to suppose that Harvey, as a recently appointed lecturer composing a course of lectures, should have had a special interest in the heart. It was, to be sure, an important organ of the body in the normal run of things, the fundamental organ for the Aristotelians and one of the three fundamental organs for the Galenists [together with the liver and brain].
>
> (p. 72)

French does not directly answer this question that he raises.

For myself I think that the work Harvey undertook, and which unexpectedly resulted in him discovering the circulation of the blood, was a sub-set of his research on the generation of animals. We recall from the previous chapter how his work over the years on the generation of animals called him back time after time to the significance of the blood and its role in the body, all of it seemingly starting with his observation of the *punctum rubrum saliens*, the capering bloody point.

So in my view, Harvey begins in the London of the early 1600s by investigating the generation of animals, and the business about the heart and arteries arose in his mind as he progressed with this. This would seem to be confirmed by the historian Arthur William Meyer, writing about Harvey's book on the generation of animals (1651), and remarking how it continues or repeats material from the much earlier book on the motion of the heart and blood in animals (1628). 'Several identical ideas are expressed in both treatises', Meyer writes, 'and sometimes in almost identical language' (p. 4). For instance, in Chapter 4 (p. 20) of the earlier book Meyer points out that Harvey writes:

> In a Hens egg I shewed [to some of my special friends] the first beginning of the Chick, like a little cloud, by putting an egg off which the shell was taken, into water warm and clear, in the midst of which cloud there was a point of blood which did beat, so little, that when it was contracted it disappeared, and vanish'd out of our sight, and in its dilatation, shew'd it self again, red, and small, as the point of a needle; insomuch as betwixt being seen, & not being seen, as it were betwixt being, and not being, it did represent a beating, and the beginning of life.

Again, says Meyer, writing about the right auricle of the heart being the last part to die, Harvey says:

> A thing of the like nature, in the first generation of a living creature most evidently appears in a hens egg within seven days after her sitting, first of all there is in it a drop of blood, which moves, as *Aristotle* likewise observ'd, which receiving encrease, and the *Chicken* being formd in part, the *ears* of the *heart* are fashioned, which beating there is always life ...
>
> (Ch. 4, pp. 16–17)

Similarly, problems about generation, including the sequence, point and so on, which Harvey writes about at the end of the book on the motion of the heart and blood in animals (Ch. 16, pp. 90–1), are taken up in his much later book on the generation of animals. Indeed, as a modern reader once I began to notice the repeated appearance of observations about the generation of animals, and especially chickens, I could see that Harvey's work on the motion of the heart and blood clearly emerges from a larger and long-term study on the generation of animals, where the first step is to identify what is happening when the heart/blood first appears. We have shown Harvey investigating the vegetable soul in our previous chapter, so here again we see him, in his research about the motion of the heart and blood, dealing ultimately with the same topic, the vegetable soul.

I think that Harvey was engaged on this research on generation – which, as he himself recognised, requires little equipment except a supply of chickens and their fertilised eggs and a sharp eye – as soon as he came home from Padua. He did not yet have the pertinent works of Fabricius to help him, for Fabricius' book on the formation of the egg and chicken (*De formation ovi et pulli*) only appeared posthumously in 1621. But Fabricius may have been Harvey's inspiration for this topic when in Padua, because Whitteridge says:

once, in Padua, in the spring season at the beginning of the century, he had helped Fabricius and had himself made observations on a cock and two hens "that I might have some knowledge of the time during which coition is most successful and the necessity therefore".[4]

In the meantime, he had Aristotle to guide him here: *Parts of Animals* III iv–vi, 665 a–668 b; *History of Animals* III xix–xix, 520 b–521 b; *Generation of Animals* extensively. Aristotle is concerned with what part comes first in development, and his general conclusion is that it is the heart in blooded animals, and the blood is also present from the beginning. In *History of Animals*, Aristotle writes: 'Now follows a description of the Blood. In all blooded animals this is the most indispensable and most universal part; it is not something acquired later, but is present from the outset in all' (HA III, xix, 520 b 10). Further, Aristotle writes:

> The blood in animals pulsates in the blood-vessels all over the body at once; and blood is the only fluid which remains throughout the whole body, and throughout life so long as it lasts. Further, blood is formed first in the heart, even before the body as a whole becomes articulated
>
> (HA III, xix, 521 a 8)

In the *Generation of Animals*, Aristotle writes: 'in all blooded animals it is the heart which can first be seen as something distinct' (740 a 20; 741 b 20); and '[in the egg] once the heart has been formed (this comes first of all) and the Great Blood-vessel has been marked off from it, two umbilical cords extend from this blood-vessel …' (753 b 20). For Aristotle the heart is a reservoir of blood (like a lake), and the two great blood vessels, the Aorta and 'the Great Blood Vessel' (the vena cava) both distribute blood as nutriment to the body. Aristotle said that 'the heart is like a living creature inside the body that contains it' (PA 666 b 20). It moves constantly and has a beat; the arteries too have a beat: the pulse.

We can perhaps envisage Harvey, opening his hen's eggs regularly, trying to spot 'what is constituted *first*, and what *last*, in a most miraculous order' (*On the Generation of Animals*, Preface, p. 20), and in time being able to recognise that the capering bloody point which is formed first is not the heart – as Aristotle said – but the blood.

Then, in my view, all this essentially Aristotelian knowledge about the generation of animals which was being acquired by Harvey, came to be complemented for him by his appointment to lecture on the anatomy of the human at the College of Physicians. If I am correct, then Harvey would be taking a sharpened alertness to issues of the heart and blood when he came to the anatomy of the human heart in his lectures. And that, I think, is the answer to the question why Harvey pays such attention to the heart.

Once Harvey came to be involved in his project on the movement of the heart and of the arteries he would investigate a whole variety of hearts.

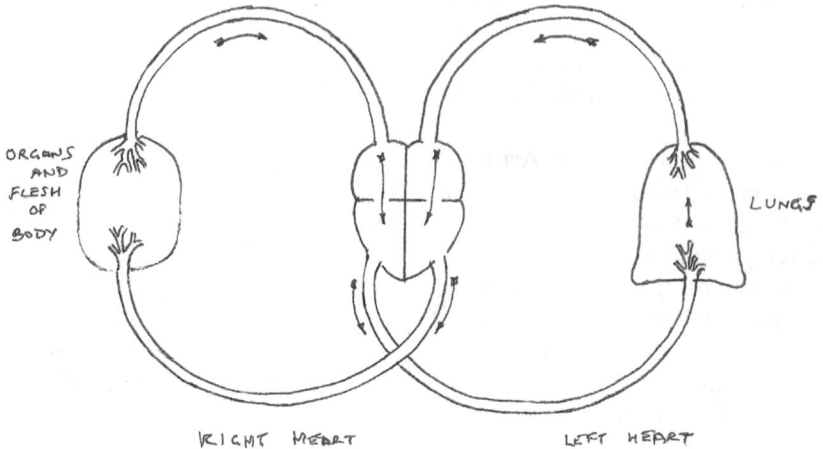

*Figure 6.1* Sketch of the circulation of the blood in an animal with four-chambered heart, such as the human. Drawing by David Lindsay Lee.

In the human heart there are four chambers. There are two on the top, left and right, which were called either 'auricles', that is ears (Harvey calls them 'deaf ears'), or 'atria' (singular 'atrium'), that is entrance rooms, as in a Roman house. Below each of these is a larger chamber, called a 'ventricle', meaning a hollow chamber like the stomach. That makes up the four-chambered heart, which is the most complex form of heart (Figure 6.1).

Perhaps confusingly, the top of the heart, as it is found in an upright human, is referred by anatomists as the 'base', because when it is cut out of the body it will stand on that end on the anatomist's bench, but not on the other end, especially because the other end has a 'tip', which moves when the heart is functioning, and is referred to as the apex.

Several kinds of animal have four-chambered hearts like the human. All animals which suckle their young have a four-chambered heart connected to a lung. Today we would classify them as mammals, though this is a category not available to Harvey.[5] They include primates such as monkeys and apes, and also cats and dogs, rats and mice, sheep, cattle, horses, goats, whales and dolphins. All birds also have a four-chambered heart connected to a lung.

Fish, by contrast, have one auricle and one ventricle, and they have gills rather than lungs for respiration (Figure 6.2).

A lot of animals have a different set-up, consisting of two auricles and one ventricle, including shell-fish, molluscs, lizards, frogs and snakes, with varying structures for respiration (Figure 6.3).

And then there are insects, most of which have one long tube rather than a heart (Figure 6.4).

These, too, would have been familiar to Harvey, because he worked extensively on the anatomy of insects and he was dismayed when his notes on them were destroyed in the course of the Civil War.

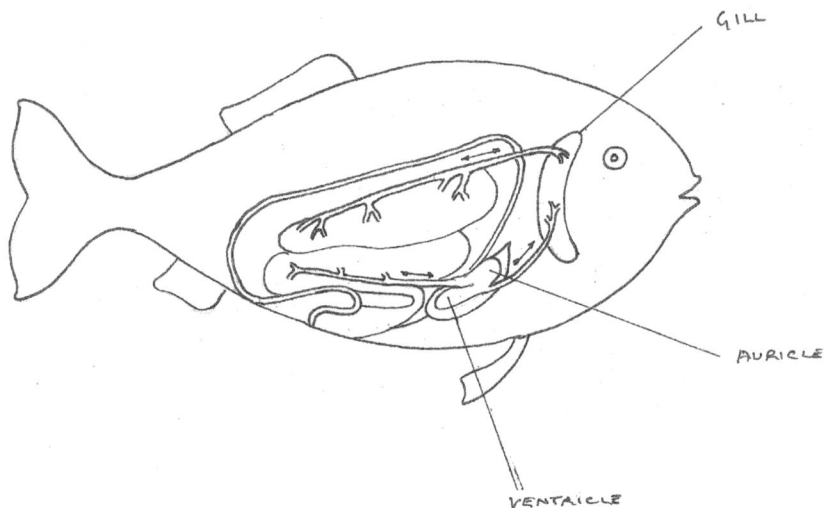

*Figure 6.2* Sketch of the circulation of the blood in a fish. Drawing by David Lindsay Lee.

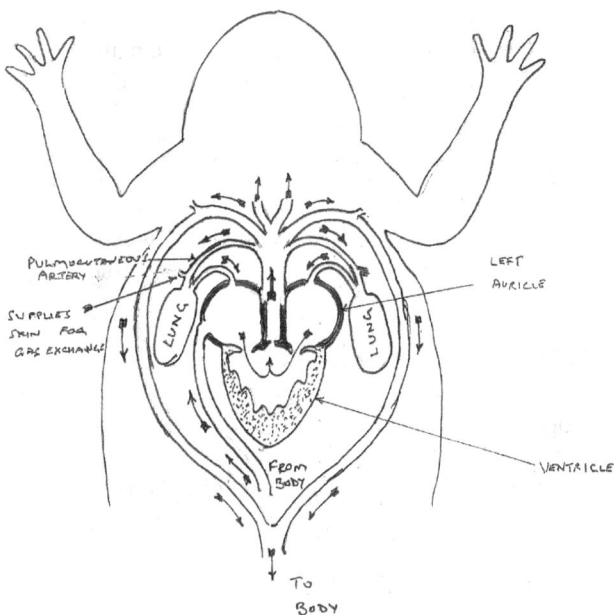

*Figure 6.3* Sketch of the circulation of the blood in a frog. Drawing by David Lindsay Lee.

Harvey of course knew all this from his own research, remarking at one point that 'there are more creatures which have no *lungs*, than there are which have, and more which have but one *ventricle* than there are which have two' (Ch. 6, p. 28). But these differences would only be noticed and remarked on

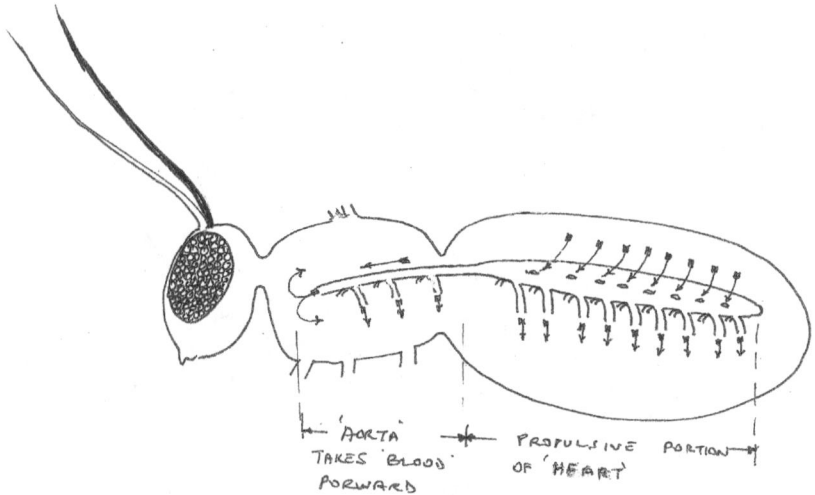

*Figure 6.4* Sketch of the circulation of the blood in an insect. Drawing by David
      Lindsay Lee.

by someone who was investigating the anatomy of a wider range of creatures
than usual, as Harvey himself was – as an Aristotelian anatomist.

    Meanwhile, there is the question of the movement of the heart, which he
found very problematic:

> When first I applyed my mind, to observation, from the many dissections
> of Living Creatures as they came to hand, that by that meanes I might find
> out the use of the motion of the *Heart* and things conducible in Creatures;
> I straightwayes found it a thing hard to be attained, and full of difficultie'.
>            [*Motion,* Chapter 1, opening sentence, p. A1v]

As he says, 'I did almost beleeve, that the motion of the Heart was known to
God alone' (A1v). In his lectures he writes, 'I have observed these things for
whole hours together, but I could not easily distinguish between them either
by sight or by touch' (*Praelectiones* f. 77r). Everything happens so fast. And
if we think about this for a moment, it means that Harvey has a live animal,
presumably tied down, he has opened its chest, if necessary by breaking
through the ribs, and now he is watching for whole hours together as the
panicked animal is dying. There is no anaesthetic, the animals are just sub-
jected to this treatment at the anatomist's will, and no-one objects in principle
to Harvey doing this. Cutting open live animals for research and demonstration had been advocated by Vesalius and Colombo in the 16th century. But
by modern standards this is a grotesque and cruel business. 'First then in the
*hearts* of all creatures', Harvey writes,

> being dissected whilst they are yet alive, opening the *breast,* and cutting
> up the *capsule,* which immediately environeth the *heart,* you may observe

that the *heart* moves sometimes, sometimes rests: and there is a time when it moves, and when it moves not.

(Ch. 2, p.4)

He is trying to determine which movement of the chambers of the heart is diastole (dilation), which is systole (contraction). As the Greek terms indicate, that there were two different movements was known to Galen, but which was which, and what did they do? One of the things he had to establish, therefore, was which is the active stroke of the heart, the stroke which moves the blood: is it systole or diastole? Traditionally, following Galen, anatomists had believed that diastole (dilation) was active and vigorous, a sucking-in of blood into the ventricle, whereas systole (contraction) was a simple relaxation. Diastole was therefore when the tip of the heart erected itself and seemed to beat against the chest wall. Harvey's experiments showed him that this opinion was the opposite of the truth.[6] One of these experiments was to use vivisection and to pierce the ventricles of the heart at these two moments, using the hearts of cold-blooded and dying animals so that he could distinguish the two moments:

> But of this [question] no man needs to make any further scruple, since upon the inflicting of a wound into the *cavitie* of the *ventricle*, upon every motion, and pulsation of the *heart*, in the very *tention*, you shall see the blood within contained to leap out.
>
> (Ch. 2, p. 6)

Harvey repeated his observations on 'many dissections of Living Creatures as they came to hand'.

Here now is the central passage in Harvey's book, describing the dramatic moment when he realised that the blood must circulate:

> Truly when I had often and seriously considered with my self, what great abundance there was, both by the dissection of living things, for experiments sake, and the opening of *arteries*, and many ways of searching, and from the Symetrie, and magnitude of the *ventricles* of the *heart*, and of the vessels which goe into it, and goe out from it, (since Nature making nothing in vain, did not allot that greatness proportionably to no purpose, to those vessels) as likewise from the continued and carefull artifice of the *doores* and *fibers*, and the rest of the *fabrick*, and from many other things; and when I had a long time considered with my self how great abundance of blood was passed through, and in how short time that transmission was done, whether or no the juice of the nourishment which we receive could furnish this or no: at last I perceived that the *veins* should be quite emptied, and the *arteries* on the other side be burst with too much intrusion of blood, unless the blood did pass back again by some way out of the *veins* into the *arteries*, and return in to the *right ventricle* of the heart.

> I began to bethink my self if it might not have a *circular motion*, which afterwards I found true …
>
> (Ch. 8, pp. 44–45)

That is as marvellous a description of a moment of discovery as any I know, and the excitement is still there. It can also be seen how many repeated observations and experiments that moment was built on.

The accounts of the stages of Harvey's discovery are covered very well by Whitteridge and by French, and do not need to be rehearsed by me here. Both Whitteridge and French build their stories about how Harvey discovered the circulation of the blood from envisaging Harvey thinking about things during the preparation for and delivery of his lectures.[7] As French puts it: 'The lectures, vivisection and disputation are the three keys to Harvey's discovery' (p. 87). Both authors emphasise the very clear steps of argument and defence that Harvey lays out in his presentation, both of them arguing that this mirrors how Harvey tried to convince his fellow physicians and special friends, along the lines of a formal disputation. Indeed, all three of Harvey's books are titled 'exercises' (*exercitationes*), the sort of exercise in disputation that lay at the centre of university education. French points out that the 1589 official statutes of the university of Padua say 'Every day teaches how much profit there is in *exercitatio*' (p. 59).

There is just one point I wish to clarify: the role of measurement in Harvey's thinking. Given the nature of the discovery, all about movement and timings, with great stress on experiment and seeing for oneself, some modern commentators have seen Harvey's discovery as essentially an instance of the 'mechanical philosophy' being widely adopted in the early 17th century. But was Harvey really inclined towards the new movement to measure things, and away from his qualitative Aristotelian ways of thinking? In Chapter 9 of his book he lays out three things to be confirmed to prove the fact of the circulation of the blood. His first argument is about the quantity of blood which must pass through the heart, and showing that it is far too much to be the product of the food we eat. 'Let us suppose how much blood the *left ventricle* contains in its *dilatation* when its full, either by our thought or experiment' (p. 49). How many drams?

> So let us imagine, that in a Man there is sent forth in every pulse of the *heart*, an ounce and a half, or three drams … The *heart* in one half hour makes above a thousand pulses … So our account being almost layd, according to which we may guesse the quantity of blood which is transmitted, counting the pulsations … But howsoever, though the blood pass through the *heart* and *lungs*, in the least quantitie that may be, it is conveyd in far greater abundance into the *arteries*, and the whole body, than it is possible that it could be supplyed by juice of nourishment which we receive, unless there were a regress made by its circuition
>
> (pp. 49–52)

Harvey says that he has tried it in a sheep, and he certainly had a minute watch with which he made his experiments,[8] but this is still clearly Harvey presenting an argument for persuasion, 'let us suppose', 'let us imagine', 'we may guess', rather than putting forward a quantified experiment as such. It is a thought experiment.

'I follow Aristotle', Harvey wrote in his book on the generation of animals. Have I succeeded in establishing my thesis that the discovery of the circulation of the blood by Harvey was part of an Aristotelian project, rather than an early instance of the mechanical philosophy of the 17th century? No-one before Harvey, with the great exception of Aristotle himself, had investigated the heart *of all animals* before, *without looking at the lungs*. Harvey's account of the motion of the heart and blood accounts for all hearts: the heart functions to move the blood from the veins to the arteries – that is what it is for, its *cause* – through all the different set-ups of ventricles and auricles of all species of animal.

Finally, the blood. Harvey specified in his book three times what his initial project of investigation was: the motion of the heart and arteries in animals (see above). Yet the title he gave to his book was on the motion of the heart and *blood*. This change of topic has not been remarked on before to my knowledge, either by Harvey or any subsequent commentator. I think there are two points here. The first is that when he started on this project Harvey was dealing with Galen's claim that the movement of the arteries (their pulsation) was different from the movements of the heart, that somehow the pulsation ran down the arterial coats, and was quite distinct from the beat of the heart. Harvey found that this was not the case, and he was even able to show that one of Galen's experiments on ligaturing the artery to show that the pulse ran along the arterial coat was impossible to perform. His own finding was that the 'two' pulses are one. As he wrote: 'the pulse of the *arteries* is nothing but the impulsion of the blood into the *arteries*' (Ch. 3, p. 12). So the things that are moving are the heart and the blood, and Harvey realised that in fact the role, the function, of the heart is to move the blood into the arteries, and on its return, to move it from the veins (the vena cava) across the heart to the arteries, again and again. The second point about why Harvey titled the book *On the Motion of the Heart and Blood in Living Creatures*, may lie with what he was repeatedly finding in his work on the generation of animals: the blood is the first thing to appear in the egg, and that the heart itself is made by the blood, not the other way round.

## Conclusions

The distinctive features of Harvey's account of the movement of the heart and arteries are that *all* the blood in the body is pumped around the body rapidly and continuously by the force of the heart, outward from the heart through the arteries, and *returned* to the heart through the veins. In this Harveian understanding of circulation, all the blood constitutes *one system*, not two, and the two types of blood vessel, the arteries and the veins, are parts of *one blood transport system*, and not two. Moreover, this concept of *unidirectional* circulation is built on a particular understanding of the functions of the flaps in the heart and in the veins, that is, that they are *valves* and that they are competent in their functioning. All these features together are definitive of Harvey's concept of the circulation, and every one of them was new with Harvey.

I have long been in two minds (as Harvey might have said) as to whether I should have included here a modern schema of the human heart as it is known today, so that one could see where the blood flows, where the valves are, and the relation of the heart in man to the lungs, and hence more easily understand the complexities of Harvey's discovery of the circulation of the blood. But there are several problems with this. The first is that the structure of the mammalian heart is extremely complex, and does not lend itself to being portrayed in two dimensions. The second is that Harvey himself does not use visual illustrations. The one visual image he commits to print, to illustrate the valves in the veins, he borrowed from Fabricius – whilst reinterpreting it. Despite all his urgings to others to use and trust their eyes, Harvey does not resort to printed illustrations: he means 'go and look at the thing itself'. His teacher Fabricius, by contrast, used many illustrations.

The third problem is the hardest to reconcile. James Shaw published an article in 1972 on 'cardiac structure and function in Aristotle', in which he had to deal with Aristotle's supposed mistakes, such as saying the heart has three chambers, where we today think it has two chambers, viz. the ventricles. Shaw points out that if, like Aristotle, you come to dissect a mammalian heart *without* already knowing its structure or function, then there is not much about its structure or function which is self-evident, except that there is a lot of blood there, which Aristotle saw as a store of blood. Shaw points out that when *we* come to look at the mammalian heart today, we *already* know its structure and function. In particular we know that there is one-way flow of blood in and out of it, that it acts as a pump to circulate the blood round the body, and so on, so we can readily 'see' its general structure because we already know its function, and also the structure and the function of its parts. Harvey had been taught the conventional Galenic doctrines about the functions of the heart, and hence of its structure. Because for Galen the arteries had to be kept topped up with a little blood from the veins, so in his account there *had to be* little, virtually invisible, pores in the septum – the wall, as it were, between the right and left ventricle – to allow some blood to seep through to the arteries. Some dissecting anatomists in the 16th century said these pores could not be found, and therefore they did not exist, but the theory still needed them. When we come to Harvey himself, he inherits from his education a Galenic view of the heart's structure and functions. As he works, experiments, dissects and vivisects so many animals, so he comes to question the Galenic view. About the pores in the septum he exclaimed: 'By my troth there are no such pores, nor can they be demonstrated'. (Proeme) As he pursues his practical research he attributed new functions to some of the structures of the heart and its appendages. The little flaps at the entry and exit of the heart became for him competent valves, and indeed he was the first to see that the chambers of the heart have fixed entrances and exits because there is unidirectional flow of blood through them. The traditionally named 'vein-which-looks-like-an-artery', carrying blood from the right ventricle to the lung, becomes in his eyes simply an artery (because it has a pulse). Similarly, with the 'artery-which-looks-like-a-vein' carrying blood from the lung to the left ear of the heart becomes for him a vein (because it does not have a pulse). All these conclusions were built on the grounds of function – function experimentally investigated. Again, Harvey redefines, from experiment, which movements of the heart's chambers are systole and which are diastole. There could hardly be a larger conceptual change about the use and function of the heart than this.

So one could say that for Harvey, as he works over the years, his concept of the structure and function of heart is somewhat fluid, with him crediting different parts with new functions at different times. You might call it a work in progress. Only when he is ready to publish, one might say, has the new function he attributes to the heart settled in his mind what its structure and the structure of its parts must be. And that final schema, if one could call it that, is what we inherit today. It is perhaps ironical that the one image of the human heart drawn to illustrate the circulation of the blood that I have included in this book is not from Harvey but from Descartes' book *On Man* (see Chapter 8 in this book), and it is one which Descartes himself might have objected to as not being 'mechanical' enough.[9]

## Notes

1 Harvey, *On Generation*, p. 283.
2 Harvey, *To Riolan 1*, p 12. Harvey is talking about motion '*in the* meseraicks [meseraic veins] *from the* intestines *to the* Liver'.
3 Peck's translation in the Loeb edition has: 'Here is an example of the method of exposition. We point out that although Respiration takes place for such and such a *purpose*, any one stage of the process follows upon the others *by necessity*'.
4 Whitteridge, *William Harvey*, 1971, p. 210, giving p. 19 of Harvey's *Generation* as her source. I would of course love this anecdote to be true, but unfortunately I cannot find it there. However, what I do find, on p. 33 of *Generation*, is Harvey writing, 'And yet once (that Fabricius may have some patronage) in the Spring time (attempting some discovery of the *time* wherein *coition* is most *succeßful*, and the *necessity* thereof) I did separate two *hennes* from the *cocke* for *foure days* space, which in that time laid three *egges* a piece, which were as *prolifical* as the rest'. There is no mention of Padua or even of a specific year, and what Harvey says he is doing is giving the written claims of Fabricius a chance, illustrated with an experiment taken from his own experience.
5 The category 'mammal' was coined by Linnaeus, and entered English in the early 1770s, so it is not one that Harvey could be familiar with. Linnaeus also originated the category 'primate', which entered English at the same time.
6 On which see French, *William Harvey's Natural Philosophy*, 1994, pp. 74–79.
7 Whitteridge, Introduction to her 1976 translation; French, *William Harvey's Natural Philosophy*, 1994.
8 John Aubrey about the dispersal of Harvey' possessions after his death. Keynes, p. 411.
9 Wilkin, 'Figuring the Dead Descartes' (2003); she points out (pp. 44–45) that Claude Clerselier, the editor of the French edition of *L'homme*, Paris, 1664, criticised Schuly, the editor and illustrator of the Latin edition, for portraying the parts of the body in a non-machine-like way and thus not in accord with Descartes' intentions.

# 7 Method and experiment

our senses ought to assure us whether such things be false or true, and not our reason, **ocular testimony, and no contemplation**.

[*To Riolan* 2, p. 73]

## Method

In his last book, *On the Generation of Animals*, published in Latin in 1651 and in English in 1653, William Harvey has an extensive preliminary section on method and experience/experiment – that is on the method of acquiring knowledge, and on the role of experience/experiment in acquiring factual information about nature. I believe that if we explore his sentiments here we will be better able to understand his use of method and experiment in his investigation not only of the generation of animals, but also in the work which had earlier led him to discover the circulation of the blood.

This account by Harvey is entirely taken up with the philosophy of Aristotle. As I remarked earlier, in his first book Harvey barely mentions Aristotle, but here now 25 or so years later he writes explicitly and loudly about Aristotle and with great admiration, urging his readers to follow Aristotle's example. Thus we, Harvey's historians, have tended to think that when writing about his discovery of the circulation of the blood Harvey was a 'modern' to whom Aristotle was almost irrelevant, yet that Harvey suffered some collapse back towards Aristotelian ways of thinking in his old age. And in a way this conclusion chimes with our view, as historians of Harvey, that the first book (seemingly without Aristotle) contains a monumental truth, whilst the other one (stuffed full of Aristotle) is of little consequence then or now.

I hope that by now it is evident that I believe that Harvey was an Aristotelian in his anatomical investigations from the very beginning to the very end, and that this is what underlay his discovery of the circulation of the blood, as well as all his other investigative work in anatomy. To put it most directly, in my view only an Aristotelian anatomist could have discovered the circulation of the blood.

But had something happened in Harvey's world such that now, 25 years or so later, he thinks it necessary to re-emphasise his dedication to Aristotle and

DOI: 10.4324/9781003247616-9

to Aristotle's method of acquiring knowledge? I believe something very important did happen in the interval between the two books which led him to speak so explicitly, expansively and admiringly about Aristotle in his later book, urging others to follow the path chalked by Aristotle.

In a word, what had happened was that in the course of Harvey's lifetime new philosophies had been proposed, become fashionable and been applied to questions of finding out about nature. When Harvey was young, Aristotle still ruled the schools and, as we have seen in Padua especially, the most for-ward-looking of all universities, there was excited fresh development of Aristotle's own approaches, in philosophy generally and especially in anat-omy. But now, in 1649–50, when Harvey handed over the text of the book on the generation of animals to Sir George Ent, not only was the political world in chaos in Harvey's eyes, but the philosophical world was also upside down. Things had moved on, including fashions in philosophy. It's a phenomenon familiar to all academics and researchers as they age: they find their topics and research looking old-fashioned and dismissed as irrelevant by the younger generation. They still feel as relevant as they ever did, and they see the younger generation as misguided, obsessed with fashion and maybe a little ignorant and arrogant. We can see Harvey as such an elderly researcher, for he was now in his early 70s. As John Aubrey recorded, Harvey 'did call the neoteriques shitt-breeches'. This expression likens them to little children who still defecate in their clothes. 'Go to the fountain-head', he advised Aubrey, who briefly wanted to study medicine, 'read Aristotle, Cicero, Avicenna.'

One such new philosophy was that created by Francis Bacon, Harvey's own patient, whom Harvey would not allow to be a great philosopher, telling John Aubrey that 'he writes philosophy like a Lord Chancellor'.

But the most important of Harvey's contemporary philosophers was the Frenchman René Descartes (or des Cartes). Young René had been educated by the Jesuits, those masters of argument, at La Flèche. Their techniques of correcting religious error or for rescuing potential heretics was to challenge their opponents intellectually, and sometimes to appear to concede points in argument to them, but always trying to turn it in their desired direction. Descartes shows he learned this lesson very well.

The most significant and challenging of his books was that *On Method*, published in 1637. The full title of the work as first published in French is most revealing for our present purposes, it is *Discours de la Méthode pour bien conduire sa raison, et chercher la vérité dans les sciences* – because that is exactly what Harvey writes about in the Preface to *On the Generation of Animals*.[1]

Harvey has three sections in this Preface precisely on the topic of how to gain reliable knowledge of the world. The first is 'On the manner and order of obtaining knowledge', the second is 'Of the former matters according to Aristotle', and the third, more specific to the book, is 'On the method to be observed in the knowledge of generation'.

For the moment we would not be far wrong in assuming that Harvey is writing a corrective here to Descartes' book, not by contending with its claims directly, but by reaffirming a better, earlier, way to conduct one's reason and to search for the truth in the sciences: that is, the way of Aristotle. After all, Harvey had a highly controversial discovery behind him which had finally proved to be true and to be widely accepted, and which was built on the philosophical foundations of Aristotle, the very philosopher Descartes and others were now trying to rebut. Descartes was quite explicit, if also furtive, about his aim of rebutting Aristotle. In a letter to Mersenne, published now in English in his *Philosophical Letters*, dated 28 January 1641, Descartes says:

> I may tell you, between ourselves, that these six *Meditations* contain all the foundations of my *Physics*. But please do not tell people, for that might make it harder for supporters of Aristotle to approve them. I hope that readers will gradually get used to my principles, and recognize their truth, before they notice that they destroy the principles of Aristotle.

We know that Harvey read this book by Descartes because he went out of his way to rebut some of Descartes' claims about the circulation of the blood and the heart's movements. We don't know whether he read the *Discours de la Méthode* in the original French from 1637, or in the Latin translation from 1644, which appeared as *Specimen Philosophiae*. Certainly he called Descartes 'a most acute and ingenious man (to whom, for his honourable mentioning of my name I am much indebted)' (*To Riolan, 2*, p. 83), but his remarks are highly critical, though politely expressed.

As is well-known about this famous and highly influential book, Descartes sets up his account of his proposed method of rightly conducting one's reason and seeking truth in the sciences as a piece of almost innocent autobiography, a little life journey to enlightenment and the improvement of the knowledge of mankind. 'I have never contemplated anything higher than the reformation of my own opinions, and basing them on a foundation wholly my own' (Part 2). He had realised, he wrote, that everything he had been taught was at best uncertain in all the sciences and philosophy except in mathematics, especially in geometry. But, he reasoned, all things are possible if one uses one's own reason in a highly disciplined way. He resolved 'to conduct my thoughts in such order that, by commencing with objects the simplest and easiest to know, I might ascend by little and little, and, as it were, step by step to the knowledge of the more complex' (second rule). In time, after eight or nine years of doubting everything, so he claimed, he was able to establish as the first principle of his new philosophy the fact of his own existence: 'Je pense, donc je suis', 'I think therefore I exist', best known even today in its later Latin formulation 'Cogito ergo sum'. From here it was a short step in reasoning to confirm the existence of a Perfect Being, God, and also of the human reasoning soul.

We have now already reached the fifth part of this little book, and here Descartes very briefly outlines a whole chain of truths he had found out about the universe and the natural world, all starting from this first principle. Commentators reckon that this is a mere sketch because Descartes did not want to suffer as Galileo had done for his heretical views. But, nevertheless, he says he would deal with light, the heavens, the planets and the earth, all on the supposition of

> what would happen in a new world, if God were now to create some-where … matter to compose one … and after that did nothing more than lend his ordinary concurrence to nature, and allow her to act in accordance with the laws which he had established.

After outlining what he would have said in the book if he had felt able to publish it, Descartes then turns to the topic of most interest to us and which is his showpiece example of the exercise of his method: the motion of the heart and arteries. So we can take this account of the heart and arteries as exemplary of Descartes' method in action.

Descartes supposes that if God were now to create the body of man 'and after that did nothing more than lend his ordinary concurrence to nature, and allow her to act in accordance with the laws which he had established' (as with his supposition about the universe), then God would create a man shaped just like us, but without any kind of soul, 'beyond kindling in the heart one of those fires without light', as in fermenting hay or wine. Unfortunately – from Harvey's point of view – Descartes says God placed in this body not only no rational soul, 'nor any other principle, in room of the vegetative or sensitive soul'. (*Ni aucune autre chose pour y server d'âme végé-tante ou sensitive*). As we saw in Chapter 5, the vegetative soul (and its operations) was precisely what Harvey was investigating in most of his anatomical research, and here Descartes says it is an unnecessary fiction.

Descartes recommends that his reader get hold of the heart of some large animal with lungs, and have it dissected in front of him to familiarise himself with its cavities and large connecting vessels. Descartes has certainly done his anatomical homework here, and describes well the cavities of the heart and the vessels and he comments (following Harvey) that the 'vein which looks like an artery', and similarly 'the artery which looks like a vein' – the two vessels carrying blood from the right heart to the lungs, and from the lungs to the left heart respectively, had been misnamed. Then the interested reader who was not hitherto versed in anatomy should take notice of 'the eleven little pellicles which like so many little doors open and close the four openings which are in these two cavities'. In the original French version Descartes speaks of '*les onze petites peaux qui, comme autant de petites portes, ouvrent et ferment les quatre ouvertures qui son en ces deux concavités*'. In the later Latin version, whose translation Descartes oversaw, he uses the term 'valves' rather than 'little doors': '*undecim pelliculas, quae veluti totidem valvulae aperiunt et claudunt quatuor ostia sua orificia quae sunt in istis duobus cavis*'.

The crucial feature of these little doors or valves for Descartes is that they make the whole operation of the heart and the movement of the blood mechanical and automatic: they *automatically* ensure one way flow. The drops of blood falling into the two ventricles (in Descartes' account) expand by the heat of the heart, and automatically close the valves by which the blood came in and, again automatically, open the exit valves. This is for Descartes an example of the only trustworthy science, 'mathematics, and especially geometry'. This is how he concludes his discussion of his method applied here to the movement of the heart and arteries: it operates like a machine, like a clock.

> But lest those who are ignorant of the force of mathematical demonstrations and who are not accustomed to distinguish true reasons from mere verisimilitudes, should venture, without examination, to deny what has been said, I wish it to be considered that the motion which I have now explained follows as necessarily from the very arrangement of the parts [*organes*], which may be observed in the heart by the eye alone, and from the heat which may be felt with the fingers, and from the nature of the blood as learned from experience, as does the motion of a clock from the power, the situation, and shape of its counterweights and wheels.

So in Descartes' approach here in his *On Method*, there is no controlling soul, there is no goal-orientedness of explanation, no 'for the sake of what?' questions. Everything just follows from the structure of the machine. And the factual content of what he 'discovers' here by his method – viz. the action of the heart and the movement of the blood – had actually been discovered by someone else: Harvey!

## Experiment

It has to be said that Descartes and Harvey do have an over-arching goal in common, what Descartes describes as a desire to promote 'greater advancement in the investigation of Nature than has yet been made' [Prefatory note]. Or, as George Ent recorded Harvey's conversation:

> 'Tis true, replied He; it hath ever been the delight of my Genius, to make strict Inspection into Animals themselves: And I have constantly been of opinion, that from thence we might acquire not only the knowledge of those leß considerable Secrets of Nature; but even a certain Adumbration of that Supreme Essence, the Creator.
>
> (*Generation*, Epist. Dedicatory, A3v)

The gist of Harvey's account of the manner and order of gaining knowledge is quite simple and direct. He says that he is publishing what he has discovered about the generation of animals so that posterity may thence discern the certain and apparent truth, but also, and that chiefly too, that (by revealing the Method I use in searching into things) I might propose to studious men, a new, and (if I mistake not) a surer path to the attainment of knowledge … especially in the Secrets relating to *Natural Philosophy*.

(*Generation,* Preface p. 3)

Nature herself, he writes, 'must be our adviser; the path she chalks must be our walk: for so while we confer with our own eies, and take our rise from meaner things to higher, we shall be at length received into her Closet-secrets'. (Preface, p. 5)

Two crucial points have made their first appearance here. First, the most central role of one's eyes – one's own eyes – their use and their reliability. Second, how our knowledge proceeds in time from lower to higher, from meaner things to higher. How is reliable knowledge built up by us as individuals? Harvey resolves apparent contradictions in Aristotle about the relation of universals to singulars: 'for Science is begot by reasoning from *Universals* to *Particulars,* yet that very comprehension of *Universals* in the *Understanding*, springs from the perception of *Singulars* in our *sense*' (Preface pp. 5–6). In order to reach true knowledge it is necessary to always remember the relation of sensory evidence to our mental ideas, and to constantly refresh our senses, especially our eyes.

'Wherefore it is', Harvey writes,

> that our *judgement* erreth about phantasmes and apparitions comprised in our minds, unless *sense* give a right verdict, established upon frequent observations, and infallible experiments. For in every *Science*, be it what it will, a diligent observation is requisite, and *Sense* it self must be frequently consulted. We must not (I say) rely upon other mens experience, but on our owne; without which, no man is a proper disciple of any part of natural knowledge; nor a competent judge of what I shall deliver concerning *Generation*; for without experimentall skill in Anatomy, he will no better apprehend it, then a man *born blind* can judge of the nature and difference of *colours*; or one born *deaf*, of *Sounds*. Therefore (discreet Reader) trust nothing I say, about the *Generation of Animals*; I appeale to none but thine eyes. For since every perfect *Science* builds upon those *Principles*, which it finds out by *Sense*: we must have a special care, that by customary dissections, we be sure those *Principles* be safely grounded. If we do otherwise, we may get a tumid and floating opinion: but never a solid and infallible *knowledge*.
>
> (Preface, pp. 9–10)

Harvey called on people to seek for truth 'from the vast book of Nature through anatomical dissections and experiments' (*Generation*, Preface, p. 22). To illustrate a little the nature and roles of anatomical experiment for Harvey it will be convenient to take some of the anatomical experiments he mentions in the course of arguing for the circulation of the blood in his book of 1628 and in his responses to his most important critic, Jean Riolan, Dean of the Paris Medical Faculty. Harvey used vivisection very extensively. His own list of animals that he dissected and vivisected included hens, geese, pigeons, ducks, fishes, shell-fish, molluscs, frogs, snakes, bees, wasps, butterflies, silk-worms, sheep, goats, dogs, cats and cattle.

Because anatomy was partly a manual art, most of its operations fitted the basic understanding of 'experiment' (*experimentum*), viz. experience, for the hands and the eyes were the primary sources of sensory experience. Therefore much of what the anatomist did was experiment in this broadest of senses, seeking the nature of the structures of the parts of the body and their actions by using hand and eye, in order then to seek their uses or purposes. For instance, the anatomist found out what materials the different parts are made of, the connections of the parts with each other, the pathways of the vessels such as the nerves, and the routes taken by the various materials which sustain the body, such as blood, sweat, chyle, urine, faeces and so on. One might say that such experiments are about the geography of the body – location, connection, flow, shape (form), material and so on.

We can take a couple of typical instances in Harvey's work. In the first one, he is talking about the little flaps at the opening of the aorta from the left ventricle of the heart, and he has already established their structure and action. Now he concludes about their use or purpose:

> that those *three pointed doors* plac'd in the *Orifice* of the *Aorta* do hinder the return of the blood into the *heart*, and that nature had never ordain'd them for the best of our intralls [i.e. the heart], unless it had been for some speciall Office.
>
> (*Motion,* p. 25)

Thus, from their structure, action and position, Harvey has worked out their purpose, which is to prevent back-flow of the blood. In other words, he has discovered that together they constitute a valve.

Similarly, Harvey reports his experiments on the nature of the blood in the arteries and veins, which he undertook to find out if there was any difference:

> if you find the same blood in the *arteries* which is in the *veins*, being bound and cut up after the same manner [as Galen bound and cut them in his experiment], as I have often tryed in dead men, and in other creatures … we may likewise conclude, the *arteries* do contain the same blood which the *veins* [contain]; and nothing but the same blood.
>
> (*Motion,* Proeme, not paginated)

It is probably possible to distinguish many different kinds of more sophisticated anatomical experiment than these in Harvey's work. I will point out just one kind, which I call *decisive*, that is to say an experiment which decides between two options. It tests whether or not something is the case, and might be called an 'experiment of the crossroads', indicating to the anatomist and his audience which is the right route to take in his reasoning.[2] Such experiments were not random or feeling-the-way, but tests brought in either to make the relation of structure to action and use more clear, or to resolve various possible interpretations.

As we saw in an earlier chapter (Chapter 6), Harvey's project, which led him unexpectedly to discover the circulation of the blood, was to 'find out the use of the motion of the heart and things conducible' or, in other formulations, to investigate 'the motion, pulse, use, action and utility of the heart and arteries', or 'the motion and use of the heart, together with that of the arteries'. One of these experiments (as we saw) was to use vivisection and to pierce the ventricles of the heart, using the hearts of cold-blooded and dying animals so that he could distinguish which movement was diastole and which systole:

> But of this [question] no man needs to make any further scruple, since upon the inflicting of a wound into the *cavitie* of the *ventricle*, upon every motion, and pulsation of the *heart*, in the very *tention*, you shall see the blood within contained to leap out.
>
> (*Motion*, Ch. 2, p. 6)

This decisive experiment on living animals had now revealed which is the working stroke of the heart: viz., when it is in tension (systole). Similarly with respect to the arteries:

> cutting or piercing any *arterie* in the very *tention* of the left *ventricle* the blood is forcibly thrust out of the wound, so cutting the *arteriall vein* at the same time, and in the *tention* and contraction of the *right ventricle*, you shall see the blood to burst out forcibly from thence.
>
> (*Motion*, Ch. 3, pp. 9–10)

Such resolving experiments had to be ones which are immediately evident to the eye of the anatomist and his audience.

We can take as a second example of such 'decisive' experimentation, an anatomical experiment that Harvey offered to Riolan, to persuade him and others of the sensory reality of the circulation:

> Cutting off a long *arterie* or *vein* any body may see this evidently by sense, when he shall see the nearer part of the *vein* towards the *heart* let out no blood, but the further part pour it abundantly, and nothing but blood ... On the other side, cutting an *arterie*, but a little blood flows from the further part [i.e. further from the heart], but the nearer part shoots with a violent force mere blood, as if it were out of a spout. By which experiment is known which way the passage is in them, either this

way or that way. Besides, you'l know what swiftnesse there is in it, what sensible motion, not by little and by drops, and with what violence to boot.

(To Riolan 2, p. 50)

Again, the experiment presents directly to the senses the answers to a whole set of questions, and for Harvey, it seems, the interpretation of the experiment was non-problematical.

It is because of this extremely strong reliance that Harvey made on the direct evidence of the senses and on the anatomical experiments that produced such evidence, that two recent commentators, Andrew Wear and Roger French, have both argued that Harvey effectively claimed that *all* knowledge ultimately came from sensation, including knowledge of causes.[3] 'If you will examine or try' whether things have been said right or wrong, Harvey wrote,

you must bring them to the test of sense, and confirm, and establish them by the judgement of sense, where, if there be any thing feignd or not, sure it will appear. Whence *Plato* sayes in his *Critias*, That the explication of those things is not hard, of which we can come to the experiment, nor are those auditors fit for Science that have no experience.

(To Riolan 2, pp. 69–70)

Therefore for Harvey, only by taking observation, dissection, vivisection and anatomical experiments together, could one reach the understanding of causes in the human and animal body. In his unpublished anatomy lectures Harvey said that the goal of anatomy was 'to know or become familiar with the parts of the body and to know them by their causes, and to know these causes in every animal "for the sake of what" and "on account of which ...".' To know 'for the sake of what' (*propter quid*) meant investigating the actions and uses of the parts.[4] Similarly, in the 1640s Harvey wrote to Jean Riolan, his most important opponent with respect to the circulation of the blood:

this is that I did endeavour to relate and lay open by my observations and experiments, and not to demonstrate by causes and probable principles, but to confirm it by sense and experience, as by a powerfull authority, according to the rule of Anatomists.

(To Riolan 2, p. 75)

Harvey is also very insistent that physiological speculations should only follow after the full recognition of the anatomical data. Harvey says that it is probable that the nutriment and other things in the blood are most likely not returned every time to the heart. But of this,

and a great many other things which are to be determined and declar'd in their proper places, to wit, in Physiologie, and the rest of the parts of

Physick, it is not fit to dispute, nor yet of the consequences of the Circulation of the blood, nor the conveniences nor inconveniences of it, before the Circulation itself be established for granted.

The example of Astronomie is not here to be followed, where only from appearances, and such a thing that may be, the causes, and why such a thing should be, comes to be, comes to be enquir'd after. But as one desiring to know the cause of the Eclipse, ought to be plac'd above the Moon, that by his sense he might found out the cause, not by reasoning of things sensible, in things which come under the notion of the sense, no surer demonstration can be to gain beleef, than ocular testimony.

(To Riolan, 2, p. 57)

Causes can be found out by the senses: there is no better persuasive than ocular testimony. This is a point Harvey repeatedly returns to. In arguing that the ventricles of the heart would be stuffed with blood, if it were not to pass into the arteries, Harvey writes: 'This consequence of mine is demonstrative and true, and followes of necessity, if the premises be true; but our senses ought to assure us whether such things be false or true, and not our reason, ocular testimony, and no contemplation.' (*To Riolan 2*, p. 73).

What is remarkable here for the present-day historian of science and medicine is that not only is Harvey's approach experimental in a modern sense as well as a 17th-century one, but that he sees Aristotle as his model for experimenting. Our current interpretation of Aristotle is certainly not as an experimenter. And although we might certainly see Aristotle as an observer, and a very good one at that, especially in the animal books, we have learned to see him as a theoretician rather than an experimenter. Yet Harvey and his teacher and guide Fabricius most certainly saw him as an experimenter. The research on animals that Aristotle performed was most evidently experimental in his work on the generation of animals, a topic on which both Fabricius and Harvey followed him.

## Notes

1 The relationship between Harvey's and Descartes' views on the heart, the blood and circulation have often been investigated by modern scholars, especially the way in which Descartes 'mechanises' Harvey's views of the heart's action, but I cannot recall anyone looking at Harvey's discussion of method here as a direct response to Descartes' views on method.

2 I use this expression from the *experimentum crucis* advocated by Francis Bacon. However, this is my terminology, not Harvey's. Harvey was not at all indebted to Francis Bacon in his experimental practice. The two men were acquainted, and Harvey is recorded by John Aubrey as saying that Bacon 'writes philosophy like a Lord Chancellor', which was not a compliment. See Keynes, *Life of William Harvey*, p. 160.

3 Wear, A. 'Harvey and the "Way of the Anatomists"' (1983); French, *William Harvey's Natural Philosophy* (1994).

4 *Praelectiones*: 'Quoniam finis Anatomiae est scire vel cognoscere partes et scire per causas et hae in omnibus Animalibus cuius gratia et propter quid ergo / Propter quid 1. Actio 2. Usus.' Harvey, *Praelectiones*, p. 6. See also p. 4: 'Anatomiae enim finis partis Cognitio propter quid …'.

# 8 'The anatomy of the blood'
## The blood as a new research object

> We use, as persons that neglect the things themselves, to pay much reverence to the specious names. The *blood* which is still at hand, and daily in our view, makes no great noise in our ears; but at the magnificent name of *Spirits*, and of an *Innate Heat*, we are strangely amused.
>
> (Harvey, *Generation*, Ex. 71, p. 460)

We have now seen that William Harvey's major discovery has suddenly made blood important, and in a totally new way: it is now seen to be one system, not two, to be very fast-moving, making rapid and continuous transits around the whole body, as well as being clearly crucial (in some still mysterious ways) to the life and functioning of the human and animal body. The different movements of the heart itself have been isolated, sequenced and identified. The blood in the arteries and the blood in the veins is now seen to be the same blood. In the wake of Harvey's discovery, in the next couple of generations, a number of English, Irish, Scottish, French, Dutch and Italian investigators and theorists all found themselves attracted to blood as *a new research object*. There was a whole new series of possible roles for the blood. Thomas Willis, Sidley (or Sedleian), Professor of Natural Philosophy in the University of Oxford, pursuing such work in his own particular way, spoke of undertaking 'the anatomy of the blood'.[1]

The blood was now thought to be something which could be anatomised. But what was blood thought to be before Harvey's discovery? I shall discuss this in several ways.

In the first place, of course blood has long had, and continues to have, *great cultural and social* importance in western culture, and possibly in all human societies. Culturally speaking, blood is everywhere. Goethe has Mephistopheles say to Faust, 'Blood is an entirely special juice', and then the two of them use blood to sign a contract, a contract which concerns the fate of Faust's soul. Blood represents – or is – life, the soul, the identity of individuals. It embodies membership in certain groups, such as the family, and the right to succession. Blood is used in religious ritual both to purify and to pollute. Christians drink what is supposedly the blood of Christ in a regular

DOI: 10.4324/9781003247616-10

ritual. Ceremony and sacredness are placed around blood. As an anthropologist wrote a century ago, 'blood actually shed means mysterious soul-power let loose', for good or ill.[2] When I was in hospital myself quite recently my surgeon called blood 'a precious resource', but also 'very dangerous stuff'.

Second, in *elite western medicine*, too, blood has long been of supreme importance. Since the time of Hippocrates, the primary curative and prophylactic intervention that the professional medical man could make was to bleed – that is, to draw off a certain quantity of blood, usually (but not always) from a vein in the inner arm. Although ideally the actual administration of bleeding was in the hands of the manual practitioner – the surgeon – the need for bleeding was (again, ideally) assessed by the mental practitioner, the physician. The quality of the withdrawn blood could be inspected and its hotness assessed, as could the range of its colours as it settled into layers in a glass vessel. Its readiness to clot or form a crust, even its taste, could give the experienced medical practitioner equally valuable information about the nature of the disease, the state of the patient, and the likely outcome. One of the rationales of bleeding was that it could (according to traditional theory) remove obstructions in the body at points distant from the site of bleeding. All physicians knew that Hippocrates had said that blood had to be drawn off 'in a straight line', though they argued fiercely about what that meant. Another rationale for bleeding was that it reduced the hecticness of the blood in the course of a fever. Yet again, it was thought to be desirable to withdraw blood which was in seasonal excess and thus causing unbalance of the humours; ideally this involved a person being bled prophylatically in spring and autumn.

To illustrate the prophylactic use of bleeding, I offer the poem which goes with this 1632 French portrayal of bleeding, an engraving by Abraham Bosse (1604–1676), 'La Saignée' (Figure 8.1).

It will be clear that the patient felt better after having been bled, and patient demand for bleeding is one of the reasons the practice was continued for so long. The lady being bled addresses the surgeon thus:

> Courage, Sir, you have begun, and I'll be brave; tighten the bandage, puncture with confidence, make a good opening. Ah, that gush of blood surprises you! How phlebotomy purifies the spirits and cleanses the blood of great putrefaction! O gods, the gentle hand, the agreeable puncture! The recollection makes me smile once more. Only a little blood drawn makes me feel much better. Above all remedies I value blood-letting. I sense my vigour return anew. But if you should consider it necessary, then operate again. I have enough spirit. I shall endure as much as you wish to do.

Basically, bleeding was used at all levels of western medical practice. William Harvey himself used it regularly as a treatment on his patients, both before and after his discovery of the circulation of the blood. As he wrote in

*Figure 8.1* The joy of blood-letting, illustrated by Abraham Bosse (1604-1676), *La Saignée*. (Wellcome Images).

*Anatomical Exercitations Concerning the Generation of Living Creatures* (1651 Latin, 1653 English):

> But while I affirm the *soul* to reside first, and principally in the *blood*, I would not have any man hastily to conclude from hence, that all *Bloodletting* is dangerous, or hurtfull; or believe with the Vulgar, that as much of *blood*, so much of *life* is taken away, because *Holy-writ* placeth the *life* in the *blood*. For dayly experience shewes, that *Letting blood* is a safe cure for several *Diseases*, and the chiefest of Universal *Remedies*: because the default, or superfluity of the *blood* is the seminary of most *distempers*; and a seasonable evacuation of it, doth often rescue men from most desperate maladies, and Death itself. For look how much *blood* is according to *Art* taken away, so many years are added to the *Age*.
>
> (*Generation*, Exercise 52, p. 296)

But given that Harvey's discovery now meant that all the blood was known to be circulating rapidly through the arteries and then the veins of the body, there was no (as it were) 'local' blood which could be selected to be withdrawn for particular medical purposes: it all came from one source, the rapidly circulating blood of the arteries and veins. Despite this new knowledge of the blood, these practices of bleeding survived long after William Harvey's discovery of the circulation of the blood, and lasted well into the 19th century.

Third, with respect to anatomy, before Harvey's discovery, blood was – perhaps surprisingly – a matter of no importance at all to the *anatomist* and to anatomical understandings of how the body worked in health. Even though in the Galenic tradition the blood was taken to be either itself the bearer of life (arterial blood) or was the sustenance for life (venous blood), yet in both these senses (vital and nutritive) it was merely a pabulum, mere food, which was consumed in staying alive. Anatomists in their textbooks, as in their research reports, simply did not deal with the blood. It wasn't just that when the practitioner anatomised a human or animal body the blood was a mere impediment to his anatomising of the solid parts, spilling out or concealing structures. More than this: the blood, both kinds, was merely seen as a 'similar part': a part which, however you divide it, is the same. And it was something *contained* within more solid structures, such as the heart, the arteries and the veins. Realdo Colombo, for instance, the anatomist who discovered the so-called 'lesser circulation' – the passage of the blood in man from the right side of the heart to the left via the lungs – does not deal with blood *at all* in his book *On Anatomising* of 1559 (*De Re Anatomica*). He does not deal with the blood in his section on the liver and the veins, apart from a marginal note simply saying 'The liver is the fount of the blood', though he deals at length with the containing vessels, the veins. Similarly, in the section on the heart and arteries, he again concentrates on the organ and the vessels, being more interested in the generation of the invisible spirits that supposedly inhabit the arterial blood than in the presence of the highly visible blood itself. Similarly, Caspar Bauhin, the Swiss anatomist whose 1,000-page *Theatrum Anatomicum* (1605) was very popular, also does not deal with the blood as a topic of concern to the professional anatomist, merely mentioning, in the course of his discussion of the liver, various differing views about sanguification (that is, blood-making, thought to be the major function of the liver). This is the case with every pre-Harvey anatomist that I know of. And nor does even Harvey himself deal with the blood in his anatomical lectures.

Thus, whilst a *physician* could observe the serum of the blood and its (supposed) fibres in the bleeding-bowl, by contrast an *anatomist* could see nothing at all about the constitution of the blood. But that was primarily because he was not looking at it, since he presumed there was nothing of interest to see in it, or at least nothing of interest that was visible to the naked eye. And this of course was also true of Harvey himself in his dual roles as practising physician and investigating anatomist. So, surprising as it may seem, the blood was simply not a subject of anatomical interest: it was not (in the modern term) a possible *research object*. Harvey's discovery of the circulation of the blood was to change all this. But even so, even with the blood now centre stage for both physicians and anatomists, as far as we know Harvey himself did not do any research on the blood in the remaining 30 or so years of his life[3]: this was done by other people.

But the biggest issue or problem with talking about the identity of the blood was that it was, itself, thought to be *alive*. It was not just part of a whole live system of the living human or animal body, but it was itself alive and it gave life to all the other parts. The Bible said this was the case, and certainly Harvey thoroughly agreed. As we have seen, in his work on generation he was obsessed with which part of the animal comes into existence first and shows signs of independent life. His view, based on his many repeated experiments and observations, was that the part that came into existence first was not the heart or other substantive organ, but that it was the blood. Before the heart formed, the blood was there, a little globule of blood, and this blood winked or pulsated, and it made – 'constructed' would not be putting too strong a term on it – the rest of the blood and all the other organs of the human and animal body, and then kept them all alive. This life of the blood, or this life in the blood, made any attempt to analyse or 'anatomise' the blood problematic.

And then there is the question of the soul: is it the soul in the blood which makes the blood alive? Which soul is it – the Greek or Aristotelian one, what one might call the 'identity soul', or is it the immortal soul as understood by Christians (and if so, did animals other than humans have such a soul?). Obviously, this soul, by virtue of being soul, would be undetectable by any physical or chemical investigation, yet would have to be recognised as the most important element of the blood.

## Chemical analysis

One of the chemically-inclined physicians of the early 17th century, John Baptista Van Helmont, had to graft his chemical analysis of the blood to his belief that the soul was in the blood. At some unknown date in the early to mid-17th century, but before 1644 (the date of his death), Van Helmont sought to distinguish the components of human blood by distillation, and these experiments were detailed in his posthumous *Ortus Medicinae* (1648). This has been argued by Allen Debus, a modern-day historian of science. There is no intellectual indebtedness to Harvey here, just a development of the chemists' customary view of the microcosm and the macrocosm. Debus's account of these distillation experiments (if indeed they were practical experiments, rather than thought experiments) is clearer than that of Van Helmont himself, so I shall use it here. 'Noting the common origin of urine and venous blood' from food and drink, writes Debus, Van Helmont

> examined each in turn. From urine a volatile salt spirit is obtained which when fermented with the earth produces salt-peter. The venous blood also contains a volatile salt spirit, which is the same as the spirit of urine except that "the Spirit of the Salt of Venal Blood cureth the Falling-sickness, but the Spirit of the Salt of Urin not so".

This led Van Helmont to conclude (continues Debus) that the vital spirit itself (something not directly amenable to chemical or other forms of analysis), 'must also be saltlike as well as airy'.[4] Whether or not practical experiments were conducted, it is clear that this is basically a logical argument, working from the perceived behaviour of the blood in the functioning human, to what the components of the blood 'must' be, in chemical terms.

We can see a similar approach taken by the Oxford professor Thomas Willis, who announced that he was giving an 'anatomy of the blood' in his book on fevers of 1659. His analysis is also purely theoretical, not experimental or practical. He produces no new facts. Whilst he celebrates Harvey's discovery of the circulation of the blood as having put paid to ancient ideas of it consisting of four humours, his own account of the blood leans very heavily on what he calls the 'Principles of the Chemists', and he argues not by evidence but by analogies. The blood clearly ferments in fevers, he argues, since we then find the body hot to the touch. It follows that its fermentation can be compared to that of wine or milk. 'There are therefore in the blood as in all fermented Liquors, apt to be Fermented, very much of Water and Spirit, a mean of Salt and Sulphur, and a little of Earth' (p. 61). And within the blood the particles of the vital spirits 'always expansed, and endeavouring to fly away do move about the more thick little Bodies of the rest, wherewith they are involved, and continually detein them [i.e. keep them] in the motion of Fermentation' (p. 59). Although this was controversial in its time, Willis's 'anatomy of the blood' is simply fashionable speculation. A hostile critic of the day, Edmund de Meara, described it as 'use of new-fangled chemical terminology taken from the wine-press and the brewery mixed with old-fashioned atomism'.[5] Despite Richard Lower's spirited defence of his master Thomas Willis against this attack, I think it pretty well sums up Willis's approach here.

However, under another hat Willis did undertake much significant and original practical anatomy – the anatomy of the brain and nerves. The resulting book, *The Anatomy of the Brain* (*Cerebri anatome*), was first published in London in Latin in 1664. Willis was created M.D. at Oxford in 1660, following his appointment as Sidley (Sedleian) Professor of Natural Philosophy in that University. He recounts that among his duties in that post was teaching on 'the Offices of the Senses, both external and also internal, and of the Faculties and Affections of the Soul, as also of the Organs and various provisions of all these' (Preface). He seems to have given one course, full of hypotheses, and only then realised that that was all mere speculation, and that he really needed to look to 'Nature and ocular demonstrations'. Filled with new enthusiasm, and assisted by friends such as Richard Lower and Christopher Wren, he 'addicted myself to the opening of Heads especially, and of every kind, and to inspect as much as I was able frequently and seriously the Contents'. Here, then, Willis is pursuing anatomical dissection, and across as many animal species as he could. By the end, he writes, 'I had slain so many Victims, whole Hecatombs almost of all Animals, in the Anatomical Court' (Dedication). However, Willis acknowledges no specific debt to Harvey's example in doing this anatomising, and the pathways in the brain that he is most concerned with are those for the invisible animal spirits – the supposed vehicles of the senses and thought – rather than those for the visible blood. Moreover, in his search here for final causes of the structures of the brain he clearly has in his own mind the image of it functioning like an 'invented Machine', as he wrote in his Dedication to the Archbishop of Canterbury, just the sort of Cartesian approach so detested by Harvey himself.

Here now by contrast is a book really reporting new experiments on the blood, experiments performed by Robert Boyle in whatever room he had adopted as a laboratory, probably at Oxford in the mid-1650s. His little book *Memoirs for The Natural History of Humane Blood, Especially the Spirit of that Liquor*, was published in 1683/4. It was a long-term interest, Boyle says, but he had lost his earlier papers, so this was a relatively recent work. 'History' is being used here in the same sense as 'natural history', that is to say an account, with nothing implied or intended about change over time. Boyle says that he deals here only with extravasated blood, that is to say blood which had been just taken out of the live, healthy human body by medical bleeding 'since I had no other at command' (p. 16). This means, as Boyle says, that he cannot here deal with questions such as the process of sanguification, the circulation of the blood, nor 'Of the Chyle, *Lympha*, and other Liquors, that are suppos'd to enter and mingle with the Blood' (p. 17), or whether the so-called humours are really constituent parts of it. Nevertheless, there are plenty of experiments that he can conduct, particularly on 'the *Serum* of Humane Blood, which is a *Natural*, and the Spirit, which is a *Factitious* part of it' (Preface). 'Factitious' means made by art, that is to say separated by experiment in the laboratory, rather than by nature, as in the case of the serum, which simply separates from the fibrous blood in a bowl. Amongst his experiments are ones to assess the heat of the blood as it bleeds out of the body, using a two-ball thermoscope; he distils it using retorts, he sublimes it, burns it, turns it into a powder, mixes it with oil of vitriol (sulphuric acid) leading to white fumes and a rise in hotness, he seals up serum of blood for a year to see whether it putrifies (it doesn't), and finds on unsealing it that the air rushes out with a whoosh. Most of his experiments are based on analysis or mixing with known chemicals.

### Physiological speculations

For Harvey and others, the blood had very important *physiological* roles, physiological here meaning what is happening in the body beyond what can be perceived by sight. Physiological speculations – and until about 1800 all claims about the physiology of the human or animal body were simply that: speculations – have always attracted more public attention than the visible facts of anatomy.[6] One Frenchman embraced the doctrine of the circulation from the moment he first heard of it, and it was primarily via his theoretical ideas that the blood became a research object in practice in the hands of other people. This was our old friend René Descartes. He did not so much adopt the circulation as Harvey taught it, as co-opt it for his own purposes.

For even as he adopted the doctrine, Descartes changed it. Descartes was, to all intents and purposes, reviving ancient atomistic doctrines. In his *Discourse on Method* (1637), Descartes wanted to present the human body as acting like a machine. One of the ways in which he did this was by describing the beating of the heart as an automatic system, sending out to the body blood which itself acted in an automatic way in accomplishing all the functions of the body. It is here that he adopted – and transformed – Harvey's new doctrine. Descartes hypothesised that the heat of the heart causes the blood, entering the heart drop by drop, to instantly ferment and expand, and that this in turn causes the heart to beat and the blood to rush out into the arteries.

As Harvey's work had shown that the blood circulates constantly and rapidly, in Descartes' eyes the blood could no longer be treated as being merely a transit system for the food (veins) and vital spirit (arteries), but it must have more important physiological roles as well. So, in portraying the human body as functioning like an automatic machine, Descartes first hypothesised that the blood consisted of particles of many different shapes. All this was at the sub-visible level – it wasn't something that could be proved, even with a microscope, but equally it was not something that could be disproved either by the test of sight. It was the sort of hypothesis which could only be sustained by its fruitfulness in explaining phenomena. So having asserted this basic point, Descartes could then argue (and it was never more than an argument) that the key physiological processes of the body are carried out automatically. For these differently-shaped particles of the blood could passively filtrate through sub-visible sieves which he believed must exist in the glands and other parts of the body. If, as he proposed, all and only the particles of one particular shape pass through a particular sieve, then, having passed through, these particles now constitute a physiological fluid with new properties, such as tears, saliva, sweat, milk, and so on. In this way Descartes hoped to offer an account of how these operations take place without special 'faculties' in the organs, and without the presence of soul overseeing them. Hence, he believed, he could show that the body acted entirely automatically, and that this automatic action arose necessarily from the mere shape and arrangement of their parts, both the visible and the sub-visible parts.

Descartes himself produced images of the dissected heart, showing the places where he believed the blood bubbled and fizzled during life. However, his own drawings for his projected book *On Man* (*L'homme*) do not seem to have been included in the trunk-load of his papers which spent three days underwater in the Seine, on its way back from Stockholm to Paris after his death. So the images of the Cartesian heart that we have today were drawn by others, and were constructed by artists working from what Descartes said in his text.

Whilst Descartes wrote his book in French, it was a Latin translation of it which first appeared in print: *Renatus Des Cartes de homine*, produced by Florent Schuly, Leiden, 1662, and it was Schuly who was the artist for engravings of the Cartesian heart (Figure 8.2).[7]

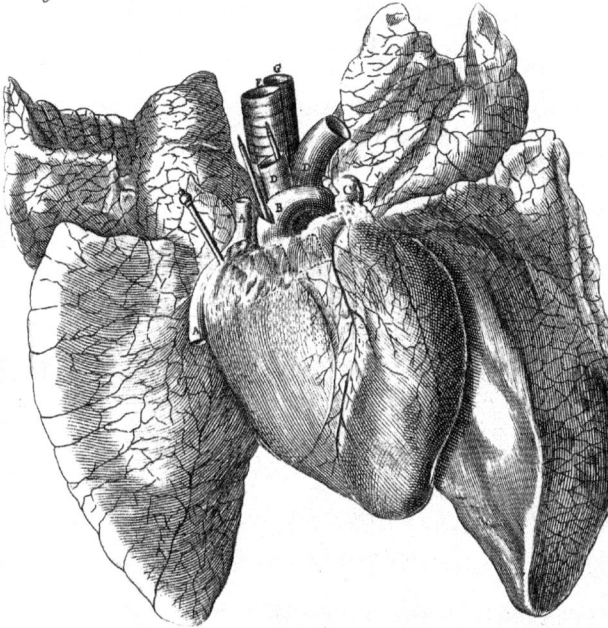

*Figure 8.2* The human heart according to Descartes, from *Renatus Des Cartes de homine*, by Florent Schuly, Leiden, 1662. (Wellcome Images).

But everything else Descartes said about the functioning of the heart was about things which could not be seen, but which were figments of his imagination.

Harvey rejected both these explanations put forward by Descartes. Fermentation or effervescence of the blood could not account for the expansion of the chambers of the heart because, Harvey said, 'there is nothing that swells so like leaven, or boyls up so suddenly in the twinkling of an eye, and falls again, but that rises leisurely, and falls suddenly', as Descartes' theory would require to happen to the blood in the heart.[8] Similarly, Harvey could not be persuaded that the mere passing of a fluid through some imagined sieve could change the nature of the fluid: 'for it does not seem probable that a fluid, by a simple and sudden filtration through the intestine, takes on another nature and forms milk', that is chyle.[9] For Harvey, Descartes was an unwanted ally.[10]

But Harvey's objections notwithstanding, Descartes' line of thinking about the blood – a line built partly, if erroneously, on Harvey's discovery – became very widely followed, and it is this which was the turning-point for blood becoming a possible research object. People did not have to become Cartesians (and there were few of these since Descartes' doctrines were thought to be atheistic), yet Descartes had now opened the door to thinking for the first time about the constitution, structure, action and use of the blood in the animal œconomy.

The dedicatory poem in the 1653 English (but not the 1651 Latin) edition of Harvey's book on the generation of animals, by 'M.LL.M.D.', demonstrates how Harvey's view of circulation could be brought together with Descartes' quite contrasting view that blood has material properties and active physiological roles in a body whose workings are strictly machine-like. The poet speaks of Harvey searching not just in dead bodies, but also searching 'in the *Living Laboratories*, when / The *Vitals* ply'd their task like *Lab'ring men ...*'

> There [viz. in live animals] thy *Observing* Eye first found the Art
> Of all the *Wheels* and *Clock-work* of the *Heart:*
> The *mystic causes* of its *Dark Estate,*
> What Pullies *Close* its *Cells,* and what *Dilate.*
> What secret Engines tune the *Pulse,* whose din
> By *Chimes without, Strikes* how things fare *within.*
> There didst thou trace the *Blood,* and first behold
> What *Dreames* mistaken Sages coin'd of *old.*
> For till thy *Pegasus* the *fountain brake,*[11]
> The *crimson Blood,* was but a *crimson Lake.*
> Which first from Thee did *Tyde* and *Motion* gaine,
> And *Veins* became its *Channel,* not its *Chaine.*
> With *Drake* and *Candish* hence thy *Bays* is curld,[12]
> Fam'd *Circulator* of the *Lesser World.*

The praise for Harvey likens him to the two first Englishmen who circumnavigated the globe, Francis Drake and Thomas Cavendish: Harvey is the 'circulator of the lesser world', that is the microcosm, man. But the poet also portrays the workings of the body in distinctly Cartesian terms, with its engines and pulleys, its wheels and clockwork.

The investigations of a group of young men in Oxford in the mid-1650s interested in the 'new philosophy' epitomise this Harveian-Cartesian view of blood and their adoption of it as a possible research object.[13] Robert Boyle was the centre of this group, and from 1656 he sought to co-ordinate their anatomical and chemical work on the blood at Oxford. We have already seen some of Boyle's experiments on blood taken out of the body, which he presumably conducted in the mid-1650s but only published about 30 years later. Among the experiments conducted in this Oxford group, most notably by the young Christopher Wren, were ones which involved using a syringe or bladder to inject milk, alcohol, medicinal drugs and other substances into live animals, in order to investigate the nature, origin and colour of the blood. These experiments are commonly referred to as 'transfusion' experiments, and it must be noted that the idea of such 'transfusion' makes no sense without the acceptance of the concept of the circulation of the blood. Harvey himself apparently thought the idea 'frivolous and completely impossible' when a similar scheme was brought to his attention in 1652 (Keynes, 313).

The idea of injecting blood into a living animal, and of doing this not by syringe or bladder, but directly from one animal to another, was first mentioned by Richard Lower in a letter to Boyle of June 1664:

> When I likewise injected many nutrient solutions, and had seen the blood of different animals mix quite well and harmoniously with various injections of wine and beer, it soon occurred to me to try if the blood of different animals would not be much more suitable and would mix without danger or conflict. And, because in shed blood (no matter how well coagulation should be guarded against by repeated shaking) the natural blending and texture of the parts must of necessity change, I thought it much more convenient to transfer the unimpaired blood of an animal, which was still alive and breathing, into another. I thought this would be more easily effected, inasmuch as the movement of blood through its vessels is so rapid and swift, that I had observed almost the whole mass of blood flow out in a few seconds, where an outlet offered. Taking hope from this, I turned mind and hands to put the matter to a practical test.[14]

Lower planned two initial experiments: one, in which two dogs of equal size should be bled 'from an artery of one into a vein of another … till they have whole changed their blood'. He does not discuss how one could tell that the two dogs had swopped all their blood. The other experiment was to bleed a large dog into a small one, to see whether the recipient would survive this operation and recover his strength. The second experiment was first tried in Oxford in August 1665, but without success. In February 1666 at Oxford, Lower, working with Thomas Willis, made his first transfusion experiment, passing the blood from an artery of each of two mastiffs (one after the other) into a vein of another dog. At the end of the experiment the donor dogs were of course dead, but the recipient seemed none the worse for it. In November 1667 Lower performed the first transfusion of blood into a human in England (the first ever had been performed in Paris in June) at the newly established Royal Society. A Cambridge University graduate, a man described as an 'eccentric scholar', one Arthur Coga, was persuaded to allow himself to be given a transfusion of sheep's blood. The experiment was deemed successful. One of the aims of this experiment had been to see what qualities might be transmitted through transfused blood. Could blood from a Quaker calm a violent man, for instance? Coga wrote an amusing letter to the Royal Society, seeking further payment, in which he described himself as 'Your Creature (for he was his own man until your Experiment transform'd him into another *species*)'. This line of research was abandoned when one of the Paris transfusees died.

Lower's own book of 1669 on the heart, the *Tractatus de Corde item De Motu et Colore Sanguinis et Chyli in eum Transitu* (*Treatise on the Heart; also On the Motion and Colour of the Blood, and of the Passage of the Chyle into the Blood*), resulted from these interests, and in particular from his vivisectional work in the winter of 1667–8, when Lower was active at the Royal

Society and engaged in experiments on respiration. In *De Corde* Lower built on Harvey's discovery of the circulation of the blood, and made a number of new and important findings. In the first place he proved that the heart was a muscle, and he showed what the structure of that muscle was. Harvey thought the heart was a muscle, but most people thought the heart consisted of a special material, with a unique pulsative virtue, on account of which it pulsated. Now it is shown to be a muscle like any other, though very complex. Whatever it is that causes it to beat and to expel the blood must now be sought somewhere other than in the specific nature of the substance of the heart. Harvey himself had written that the motion of the heart 'was like that of the *muscles*, where the *contraction* is made according to the drawing of the *nervous parts, and fibers,* for the *muscles* whilst they are in motion, and in action, are envigorated, and stretched, of soft become hard, they are uplifted, and thickned, so likewise the *heart*' (Ch. 2, p. 6). Indeed it was two more centuries before an investigator, one Dr Robert Lee, working in London, was able to discern the nerves and their ganglia in the flesh of the heart muscle.[15]

With respect to the origin of the heat of the blood, a much-disputed question of the time, Lower showed that the heart was not the source of the heat of the blood, and neither did any chemical transformation occur in the heart to produce the heat of the blood. He showed that the change in colour of the blood (from the dark red of the veins to the bright red of the arteries) takes place in the lungs, and not in the heart. Most important of all, he showed that this change of colour in the blood occurs through the air (or some component of air, which he sometimes referred to as the 'nitrous spirit') actually entering the blood during respiration, and that this entry of air into the blood is essential to the life of animals.

Soon there were other investigators who wanted to explore the structure and nature of the blood. The Dutchman Antoni van Leeuwenhoek, for instance, who linked himself to the Royal Society investigators by consistently sending his findings to London to be published in the *Philosophical Transactions*, reported on his first microscopical findings on the blood in 1673:

> I have divers times endeavoured to see and to know, what parts the *Blood* consists of; and at length I have observ'd taking some Blood out of my own hand, that it consists of small round globuls driven through a Crystalline humidity or water: Yet, whether all Blood be such, I doubt. And exhibiting my Blood to my self in very small parcels, the globuls yielded very little colour.[16]

The structure of the blood remained a research interest for Leeuwenhoek for many years. Together the microscope and chemical analysis were beginning to produce an experimental 'anatomy of the blood', and to show that it has structure and constituents, as well as life and nourishment.

## Notes

1  Thomas Willis, *De febribus* (1659), p. 1.
2  Robinson, 'Blood' (1909), p. 715.
3  His main discussion of blood is in Exercise 52 of *Generation,* but it is a theoretical discussion not an investigative one.
4  Debus, *The Chemical Philosophy* (1977), vol. 2, pp. 366–367, citing Van Helmont's *Oriatrike* (1662), an English translation of his *Ortus medicinae* of 1648.
5  Edmund de Meara, *Examen Diatribae Thomae Willisii De Febribus*, London, 1665, as translated in Dewhurst, *Richard Lower's* Vindicatio (1983), pp. xxiii–xxiv.
6  On this see Cunningham, 'The Pen and the Sword' (2002).
7  Wilkin, 'Figuring the dead Descartes' (2003).
8  Harvey, *To Riolan 2*, p. 84.
9  Harvey, *The Third Letter. In Reply to Robert Morison, M.D. of Paris*, pp.196–197.
10  On these issues see now Fuchs, *The Mechanization of the Heart* (2001).
11  Pegasus, the mythical winged horse, 'was said to have created various springs of water from the earth by the stamp of his foot, including Hippocrene on Mt. Helicon, near the Muses sacred grove', Hornblower and Spawforth, *The Oxford Classical Dictionary*, 1996, p. 1131.
12  Bays or baies: i.e. leaves of bay in garland to reward a conqueror; hence, figuratively, fame and repute. Candish is a contraction of Cavendish.
13  The most extensive account of the enquiries of these young men is given in Frank, *Harvey and the Oxford Physiologists* (1980).
14  Lower, *Tractatus de Corde* (1669), quoted from chapter 4 (unpaginated).
15  Lee, *History of the Discoveries of the Circulation of the Blood etc.* (1865).
16  *The Philosophical Transactions of the Royal Society*, London, April 1674 (no. 102), p. 22.

# 9 Precursing Aristotle

## Why and how did we lose this Aristotle?

In this chapter I try to open the question why or how we lost our understanding of Aristotle the investigator of 'the animal', as characterised earlier in the present book.

Everything I have tried to do in this book has been built on my view that Aristotle was not a modern investigator in his animal books, but that he was engaged in a unique and very precise programme of investigation into 'the animal', a programme answering the needs of his time and his philosophy. I have also argued that, centuries later, and in a different society, first Fabricius and then Harvey were to try and resuscitate this programme of investigation. This programme of investigation is not followed anywhere today by anyone. But almost everywhere that Aristotle's animal books are translated or discussed today his work is described as either originating or foreshadowing modern disciplines, and he is described as a precursor of modern biologists and the like. If my argument is to be assessed fairly then the whole issue of precursorism needs to be discussed.

The position of precursors and forerunners in the history of ideas is very strange: they are given a role by virtue of something which happened after their death and of which they could have had no inkling. Thus, for instance, Harvey's supposed 'precursors' are defined as such only in the light of Harvey's work, which was performed long after they were dead. It is a case of what has been called 'backward causation', and it is meaningless as history. The words 'curse' and 'cursor' are quite unrelated, but nevertheless I feel that once we have made Aristotle (or whoever) into a precursor of a later person, of a later concept, or of a later discipline, we have in a way cursed him in that we have just made it impossible for ourselves to find out what he was really doing. So, whenever these days I see historians writing about 'precursors', I curse quietly to myself.

The issue is one that will not go away. It just seems so natural to seek out in the past predecessors for people and their achievements in the present, and to seek 'fathers' for modern disciplines, and it feels as if we are doing proper history of ideas when we find such candidate precursors. But to enable me to read Harvey as a happy follower of Aristotle in anatomising animals, it was necessary for me to simply disregard the modern reputations we have given to Aristotle as a precursor (good or bad) in so many areas of the study of living

DOI: 10.4324/9781003247616-11

things. So in order to make a start at lifting the curse of precursing, I will here look at how Aristotle got some of his posthumous reputations, reputations which have been standing in the way of us understanding what he was doing with respect to the animal books – and hence also of what Harvey was doing much later.

For many centuries the works of Aristotle were the stuff of the school and university curriculum, and his books were printed and reprinted in their thousands. Gradually, and especially in the 17th century, some aspects of the Aristotelian world-view were challenged, especially in the realms of cosmology and physics, by people like Galileo Galilei, Francis Bacon, René Descartes and Isaac Newton. By the end of the 17th century Aristotle had become merely an historical figure in cosmology and physics, whose writings and interpretations had been comprehensively replaced by new, modern, alternatives. It was no longer a profitable affair to print editions of Aristotle's writings. It was not until the 1790s that printers and editors again began to issue the complete works of Aristotle in print, first in Greek (Buhle, 1791–1800, at Strasburg), then in Latin (Bekker, 1831, in Berlin) – but this time primarily as an historical exercise, as documentary relics of the classical world – a world that was now seen as having been different from the present.

But in the realms of sublunar natural phenomena, and especially in the study of living things, Aristotle's writings were not rejected wholesale in the 17th century as just a relic of the classical world, nor in the next two centuries either, even though several new approaches to the study of living things were put forward in the course of the 18th and 19th centuries. It was not that Aristotle was being used as a model to follow anymore (as we saw had been the case for Harvey and a few others in the mid-17th century), but that when naturalists looked at the animal books of Aristotle, as they occasionally sometimes did, they felt that they and he were interested in the same subjects and asking the same questions, even though they and he may not have come to the same conclusions. Why have Aristotle's animal books repeatedly continued to look *familiar*, rather than *un*familiar, to scholars over the last 300 years or so? Or rather (since books are inanimate objects which don't do anything), why have *we* chosen to see them as familiar rather than unfamiliar? The point was made by George Henry Lewes in his refreshing and often hilarious discussion in *Aristotle: A Chapter from the History of Science* (1864).

> The summary treatment which sufficed in the case of the *Physics* [to persuade the modern reader to dismiss it] would, in the case of Biology, have carried no conviction. The reader was prepared to find the Physics altogether valueless, but he is told that in the department of Natural History, Aristotle made important discoveries, anticipated some of the brilliant results of modern research, and laid the "eternal bases" of the Science. Nothing but a detailed examination would suffice to elicit the truth on such points.
>
> (pp. 155–156)

But having raised the question as to why the Aristotle of the animal books looks familiar to modern people, Lewes chose not to answer it directly. Instead, Lewes set about a point-by-point demolition of all the claims that the moderns had made for Aristotle the 'biologist' and his supposed discoveries and anticipations, and in this way Lewes made Aristotle look unfamiliar.

But the question about the apparent familiarity of the Aristotle of the animal books deserves exploring, for it also underlies our modern willingness to see Aristotle as a biologist and comparative anatomist and systematist. Why do we – rightly – reject out of hand any thought that Aristotle the physicist was practising an early version of modern physics, but willingly think that there was an Aristotle the biologist (etc.) who was millenia ahead of his time? Why do we find the physics of Aristotle a relic of a distant and lost time, yet think Aristotle's animal books could have been written yesterday? Were there two completely distinct Aristotles in one human body, or two completely distinct brains in one head, or what? Why, indeed, have naturalists and biologists of the last 200 years chosen to give biology a long history as a discipline, even when they know it only has a short history?

For the modern scholar and student, translations of Aristotle's texts into modern languages outlive any particular interpretations of them, and we tend to see such translations as neutral attempts to transfer meanings from one language into another. But of course translations are, in part, unavoidably also interpretations. If we look at the views of some translators of the animal books into modern languages, we can see both the stages in which Aristotle was given his modern reputations – and thus how the animal books were given new, modern identities – and also see Aristotle himself progressively being dressed in new clothes (as it were). We can also see how translators and commentators generally wear their prejudices boldly on their sleeve as they transform whatever it was that Aristotle was doing into some 'pre-echo' (another awful term) of modern practice. We shall find that they welcomed Aristotle's animal books as instances of natural history, zoology, comparative anatomy, biology – or whatever was the umbrella discipline in the field of life which was dominant at their particular time. Translating a book is a big commitment and requires in the translator a belief that what they are doing and whom they are translating are important today. It is perhaps not surprising that translators sometimes make big claims for their heroes and for their significance today. Here I shall take some sample translators in reverse chronological order, to trace the changes in the supposed identity of Aristotle's aims and intentions in the animal books.

In 1913 D'Arcy Thompson (1860–1948), later the author of *On Growth and Form* (1917), presented a talk at Oxford called 'On Aristotle as a Biologist'.[1] Fresh from translating Aristotle's *History of Animals* into English (1910), Thompson declared that 'Aristotle seems to me to have been first and foremost a biologist, by inclination and by training' (p. 11). That is, that Aristotle was a biologist, and not only that but that he was a biologist even before he was a philosopher, or anything else. Thompson argued that this biology was not carried out in Aristotle's old age,[2] but when he was a young man and living in or near Mitylene on the Asiatic coast, before he taught Alexander and before he taught in the Lyceum. Thus, Thompson wrote, 'it follows for certain, if all this be true, that Aristotle's biological studies preceded his more strictly philosophical work; and it is of no small importance that we should be (as far as possible) assured of this, when we speculate upon the influence of his biology on his philosophy' (pp. 13–14). Thompson waxes lyrical on the similarity between Aristotle's views and those of his own day:

> his language, and his methods and his problems are wellnigh identical with our own. He had familiar knowledge of a thousand varied forms of life, of bird and beast, and plant and creeping thing. He was careful to note their least details of outward structure, and curious to probe by dissection into their parts within. He studied the metamorphoses of gnat and butterfly, and opened the bird's egg to find the mystery of incipient life in the embryo chick. He recognized great problems of biology that are still ours today, problems of heredity, of sex, of nutrition and growth, of adaptation, of the struggle for existence, of the orderly sequence of Nature's plan ... It would take more than all the time I have, to deal with any one of Aristotle's theories – of generation, for instance, or of respiration and vital heat, or those still weightier themes of variation and heredity, the central problems of biology, or again the teleological questions of adaptation and design.
>
> (pp. 14–15, 17)

D'Arcy Thompson then lists some of the more striking observations about animal structure, life and behaviour which Aristotle first noted, drawing special attention to observations which had been remade by modern observers only in very recent times. He concludes (surprise, surprise) that Aristotle's biology did have an influence on his philosophy. For instance,

> in his exhaustive accumulation and treatment of political facts, his method is that of the observer, of the scientific student, and is in the main inductive. Just as, in order to understand fishes, he gathered all kinds together, recording their forms, their structure and their habits, so he did with the Constitutions of cities and states [creating] a Natural History of Constitutions and Governments.
>
> (p. 25)

Thompson's characterisation of Aristotle as a *biologist* is somewhat unusual for this date, and has to do with the fact that this speech was the Herbert Spencer Lecture, and Thompson was celebrating Spencer as much as Aristotle. Spencer had wanted Biology (with a capital B) to be one of the departments of his great *Synthetic Philosophy*, and the two-volume work which filled this role, *The Principles of Biology*, had appeared in 1864 and 1867. Moreover, Thompson was a pupil at Cambridge of the professor of physiology (that is, of *experimental* physiology), Michael Foster, who was particularly keen to claim that the umbrella-science of the organic should be biology. Thompson's choice of 'biologist' to describe Aristotle was almost prescient, and thus has not gone out of date: rather, it has come into date. Spencer himself was not concerned at all with history in his *Principles of Biology*. Thompson is probably the first person volunteering to outline a history of biology which traced it back to Aristotle and forward to Spencer. Thus Thompson's view of biology is of a subject which has a long if intermittent history, with Aristotle as its founder. And, one might ask rhetorically, who better to recognise Aristotle as the earliest biologist than Thompson, who was himself Professor of Biology (1884–, a title later changed to Natural History) at the University of Dundee, and then Professor of Natural History at the University of St Andrews (1917–), and who had learned his Greek at the knee of his father, who had been a Professor of Greek? But since this is an instance of disciplinary history by a practitioner of that discipline, we might ask also: was Thompson attributing to a past actor, Aristotle, a position and practice held and pursued by a living actor – himself?

Moreover, Thompson's characterisation of Aristotle as a biologist is gloriously over-the-top explicit. This Aristotle saw 'problems in biology wellnigh identical to our own'; 'he recognised great problems of biology that are still ours today, problems of heredity, of sex, of nutrition and growth, of adaptation, of the struggle for existence, of the orderly sequence of Nature's plan'. In other words, looking at Aristotle through the spectacles of early 20th-century biology, Thompson saw Aristotle as a fellow worker, and in doing so lifted Aristotle, body and soul, out of the 4th century B.C. and into the early 20th century. In fact, by claiming that Aristotle was interested in problems 'of the struggle for existence, of the orderly sequence of Nature's plan', Thompson was making Aristotle into someone with the concerns of Darwin and Spencer! I trust it requires no argument to demonstrate what an absurd position this is to take if one is trying to reach the original identity of the activity of Aristotle.

Having seen Thompson in the early 1910s, now let us go one step back, to William Ogle, whose translation of *Parts of Animals* into English was first published in 1882, and is still the one preferred by many scholars.

Ogle was both a doctor and classical scholar. Ogle's Introduction begins abruptly with the words: 'How came these adaptations about, is a question coeval, we may be sure, with the first recognition of the adaptations themselves.' Well, we can actually be quite sure this was the case – but not in the way that Ogle meant it. For according to the *Oxford English Dictionary*, a historical dictionary which records first usages, 'adaptation' in the sense in which Ogle intends it here for animals as 'organic modification by which an organism or species becomes adapted to its environment' was first used in English by Charles Darwin in 1859! So, yes, as soon as Darwin and some of his contemporaries (especially French investigators) recognised differences between animals as being 'adaptations' over time, then the question of how these adaptations came about did immediately arise. But that was in the 19th century, not in antiquity as Ogle imagined. Ogle continues: 'The answers to it fell of old, as ever since, into two main divisions', a materialist one, where philosophers such as Democritus proposed that the phenomena arise from the properties of matter, and a spiritual, intelligent or teleological one, in which God or Nature is an active, foreseeing, agent, and has designed everything perfectly. 'Between these two views', Ogle claims, 'Aristotle had now to decide'. Although Ogle makes a strong case that this was, as he described it, an ancient question which had persisted for more than 20 centuries, we can be certain that it had not been. Adaptation was a word new in the 19th century, for a new concept. Unwittingly, but perfectly understandably, Ogle has transferred the Darwinian challenge of the late 19th century back to Aristotle's time. In portraying the ancient materialists Ogle even said they were 'partly anticipating Darwin' (p. ii). It is clear that Ogle is a Darwinist in his thinking, and that for him Aristotle is an early Darwin or someone close to him. So this first sentence of Ogle's Introduction is a give-away and reveals all. Ogle's Aristotle, like Ogle's Darwin, is of course portrayed as a biologist, and the animal books as his 'biological treatises'.

This interpretation of Ogle's use of the concept of 'adaptation' as placing him as a Darwinian, and of his portrayal of Aristotle as an early Darwin, is confirmed by correspondence between Ogle and Darwin themselves. This had been going on for 20 years or so by the time Ogle published his translation. He sent a copy of the book to Darwin, now an old man, with a letter in which he says, 'I feel some self-importance in thus being a kind of formal introducer of the father of naturalists to his great modern successor', whilst also calling Darwin 'Democritus come to life again'.[3] Darwin was suitably surprised and pleased by being thus introduced to Aristotle – whom he had never before read.

Whilst Aristotle is, for Ogle, a biologist of a generally 19th-century mode, he also discusses both what he took to be Aristotle's shortcomings as a biologist and some reasons for them. For instance, Aristotle's senses 'told him nothing of the origin of this [co-ordinating] force, and, whatever may have been his ideas as a metaphysician, as a biologist he was silent' (p. ix). Again, 'Aristotle never expresses himself clearly on this matter [of the organisation of an individual species]' (p. x), or

> in his search for final causes, in his attempt, that is, to assign to each organ its proper function ... it must be confessed that his success, as measured by what has been attained in modern times, was but small ... He was trying to solve the complex problems of biology, while the ancillary sciences were yet unknown.
>
> (p. xi)

Ogle also defends Aristotle against charges of blunders in identifying creatures or parts, or of hasty generalisation: 'The stage to which biology had then attained made this a matter of necessity' (p. xv). But at least Aristotle was (in Ogle's view) earnestly trying first to *collect facts* and then to *sort them*, if necessary by 'temporary generalisations', which Ogle suggests are the first two steps in science. Indeed, Ogle writes, 'it is only in contrast with the achievements [of the moderns] that we can speak of his results as failures' (p. xvii). In these ways Ogle is perhaps the most sympathetic modern reader of Aristotle's animal books – but still persisting in seeing him as a Darwinian biologist or protobiologist. His Aristotle is a man before his time – but only because he presents him as a man *after* his time, that is as Darwin.

Contemporaneously with Ogle in England, in France Jules Barthélemy-Saint Hilaire, putative son of Napoleon, produced a translation of the *History of Animals* in which he considers Aristotle to be a *zoologist* and claims that 'On all the essential points, Aristotle is comparable to the most advanced Moderns' (p. clxiii). That was in 1883. Two years later, in his translation of *Parts of Animals* Barthélemy-Saint Hilaire now regards Aristotle as a *comparative physiologist*:

> Comparative physiology, comparative anatomy! These words might perhaps appear somewhat ambitious when you hear them applied to this ancient relic. But there is no mistake: if the Greek genius did not invent the word, it invented the thing – which is the better of the two.
>
> (p. iii)

He even claims that if you take all Aristotle's 'physiological' works together, 'we don't exaggerate when we compare his "comparative physiology" to a course taught today by a member of our Institut national'. Aristotle was thus the original Georges Cuvier:

> In our 19th century Cuvier is in agreement with his predecessor: movement, sensations, digestion, circulation [*sic*!], respiration, speech, generation, secretions and excretions, these are the divisions of an excellent comparative Anatomy. Do we not recognise the divisions which Aristotle proposed?
>
> (p. vii)

Aristotle is thus not just a modern – a modern comparative anatomist – but also a Frenchman. And once Barthélemy-Saint Hilaire published his translation of *On the Generation of Animals* in 1887, he saluted the work as 'a zoological masterpiece!' and as a piece of *embryology*. 'If the word embryology is new', Barthélemy-Saint Hilaire announced,

> the science itself is not; and one can see it in full in the Aristotelian work, three centuries and more before the Christian era … It is Aristotle who created this science, just like he created so many others … Aristotle didn't stop at animals properly so-called: he foreshadowed this other science, which has only just been born among us, and which we call biology, whose object it is to study life in all organised beings, from plants up to the higher animals, including man.
>
> (pp. iv–v, vii–viii)

Both Ogle, in his somewhat sympathetic defence of Aristotle's findings and approach, and Barthélemy-Saint Hilaire, in his forceful exclamations on the modernity of Aristotle, were directly opposing the arguments and criticisms of another scholar of their day, who is our next earlier writer on Aristotle: George Henry Lewes. Lewes, the partner of the novelist George Eliot (who borrowed his first name for her *nom de plume*), was a medical man, popular science journalist, experimental physiologist and philosopher. His particular philosophical persuasion was that of August Comte, whose writings he translated into English. His extended critique of Aristotle was written as part of a projected history of philosophy. As a Comtean, Lewes saw the history of philosophy as necessarily having three stages, corresponding to the essential stages in the development of mankind's thinking. The first stage is *theological*, with most phenomena ascribed to invisible powers, spirits, deities or demons – all of them variable or capricious; it aims at, or presumes, knowledge of ultimate causes. The second stage of human knowledge is *metaphysical*: it is like the first stage but without the idea of these agencies being variable; ultimate causes are seen as inseparable from the objects. Lewes gives as an example the vegetal or vegetative soul, which I have quoted before (see Chapter 5) and which is a particularly interesting example for the case of Aristotle's animal books.

> For example the Vegetal Soul, which is supposed to be the cause of all the phenomena observed in plants, is not a plant, nor a property of the plant, nor the resultant of the plant's many properties; it is an existence *sui generis*, in virtue of which the plant *is*. At the same time this Vegetal Soul is exclusively limited to the plant; it has no other form of existence; it exists only under the conditions of plant-life ...
>
> (p. 30)

And it directs everything. Thus the ultimate cause of vegetal life is the 'vegetal soul': yet though it is immaterial it operates consistently, and without recourse to outside, hidden, agencies. This is the stage that Aristotle in his animal books had reached in humanity's progress towards knowledge, in Comtean eyes: the metaphysical stage.

Finally, the third and highest stage in the progress of human knowledge for a Comtean is the *scientific*: the guarantee of science is in the verification of experience, direct or indirect. Science's pronouncements are never final, and its practitioners never presume knowledge of ultimate causes. Their aim by contrast is to detect laws of nature.

'These three modes constitute the necessary evolution of speculative thought', wrote Lewes (p. 34). The dawn of the scientific was among the Greeks. 'It is in them that, for the first time, appears the systematic effort to ascertain the relations of things objectively, *to detect the causes of all changes as inherent in the things themselves*, and to reject all supernatural or outlying agencies' (p. 44). In other words the Greeks were at stage two, the metaphysical, which is itself the dawn of stage three, the scientific. But it was a long

time a-dawning: for it is only after another 2,000 years that there was a revolution in thinking; 'and during the two hundred years which succeeded that
revolution, almost everything we now dignify by the name of scientific truth,
saw the light' (45). So for Lewes, Aristotle is in no way a modern, practising
early versions of modern disciplines or of modern ways of thinking.

Still of value here in helping us not 'read into' (p. 200) Aristotle's animal
books our modern categories, disciplines and discoveries, is Lewes on classification. Lewes writes

> Are we justified in interpreting some of [Aristotle's] generalisations as
> profound attempts at classification? There are probably few who do not
> believe that among his claims to eminence as a biologist, must be named
> the first outline of a scientific Classification; and were this idea correct,
> his rank would indeed be very high, for Classification is one of the latest
> results of scientific research. I may say at once that it is only by bringing
> together certain general statements, and *disregarding the whole context*,
> that a plausible scheme can be drawn up from his works; and so far from
> his having laid "the eternal bases" [as Cuvier claimed] upon which mod-
> erns have erected their classification, it does not appear that he had ever
> *attempted* a special arrangement of the various groups of animals.
>
> (p. 273, emphasis as in original)

Unfortunately even today Aristotle the classifier and precursor of Linnaeus
has not disappeared, despite Lewes' justified criticism.

Barthélemy-Saint Hilaire was to counter Lewes' position very aggressively:

> these principles [of Comtean philosophy] do not aid one to properly
> judge the past of the sciences, nor to understand, as one should, the
> route that they follow in their incessant progress. To suppose gratuitously
> that science is first theologic, then that it becomes metaphysical, and that
> after these two aberrations it finally becomes positive, is to claim that
> science is very recent, and that it dates in some wise from the 19th cen-
> tury, where Positivism has finally hauled it from its wayward path.
> Nothing is less true than that hypothesis
>
> (p. xxv)

As we might expect, all translators of these books start with some image of
Aristotle's enterprise in their mind, and this guides them in their translation
and interpretation. In the case of D'Arcy Thompson, for instance, we know
that however careful, diligent, cautious he might have been as a translator,
nevertheless he already knew they were the works of a biologist. And someone who both before and after making the translation thinks of Aristotle as
a biologist is going to try their best to make him sound like one in the translation and commentary! I think we can safely say that no modern translator of
Aristotle's animal books has taken up the exercise innocently, then reached
halfway in the long and laborious exercise when they suddenly and

unexpectedly thought 'Wow, this is biology, and I never realised until now!' No: the disciplinary status of the animal books is something that translators know, or rather assume, before they start. It is, in other words, a prejudice, something judged in advance of, and then probably in disregard of, the evidence. It is of course far from being a problem limited to the animal books or to Aristotle, though it is particularly marked there.

Our next step backwards in tracing translators of Aristotle's animal books takes us to a revision of Scaliger's 16th-century translation into Latin, and published at Leipzig in 1811. Johann Gottlob Schneider, himself a naturalist who had published *Some Specimens of the Zoology of the Ancients, Taken from the Natural History of Fish* (1782) and who had worked on amphibians, now published Aristotle's *History of Animals* in Latin and Greek in Leipzig in 1811, and dedicated it to Cuvier: 'To the most learned G. Cuvier, Fellow and Secretary of the French Imperial Literary Institute, outstanding Zoologist and Zootomist, this new version of an Aristotelian work is dedicated'. So, as Barthélemy-Saint Hilaire was to do later, in his dedication Schneider believes his restored Aristotle is of relevance to the outstanding zoologist of the day, Cuvier, and that the two of them were engaged in comparable, maybe even identical, activities.[4]

So far then, in retracing the sequence of translations of Aristotle's animal books, we have found that to his translators Aristotle, when speaking of animals, has always been a modern. Either Aristotle reminded the translator of some eminent investigator of life in his own modern day; or (more likely) the work of some eminent investigator of life in the modern day seemed, to the eyes of the translator, to be visible in some earlier form in the work of Aristotle. Thus if Cuvier is the latest modern investigator whom one admires, then Cuvier is a second Aristotle, and Aristotle the original Cuvier. Two great moderns have now been compared to Aristotle, and vice versa: Darwin, and now Cuvier.

Finally in our reverse sequence we come to Aristotle in his last – that is, his earliest – modern guise, and this time as an early Georges Louis Leclerc Buffon (later Comte de Buffon), and here Buffon comes to be portrayed as the second Aristotle, and Aristotle as the first Buffon. In France Armand-Gaston Camus, a lawyer and later an enthusiastic revolutionary, published a French translation of the *History of Animals* in 1783. He began it seeing the success that had come to a translation of Pliny (by Louis Poinsinet de Sivry, 1771): wouldn't the original which Pliny copied not be even more successful, he thought? How Camus interprets Aristotle's work is fascinating. The discipline of biology was not to be invented for another quarter of a century, so was unavailable to him, even if he had wanted it, in order to characterise the *History of Animals*. Rather he sees the *History of Animals* as an instance of *natural history*, a thriving *18th-century* discipline, and thriving in several rival forms. Camus dismisses the natural history of 'the dry nomenclatures, however well arranged by orders, classes and genus', in other words, the natural history of Linnaeus and his followers. Rather he sees Aristotle more as a natural historian in the mould of Buffon. We have seen how crucial a role

Aristotle's *On the Soul* played in Aristotle's own project. But by the 18th century its role in the animal books had been virtually lost. Buffon did not see its relevance in the natural history he presented in his great sequence of volumes,[5] and Camus, too, omits it from his listing of Aristotle's animal books. But in other respects Camus describes Aristotle's project in a way which we can now see is quite close to Aristotle's own:

> The plan of *The history of animals* is grand and vast. What Aristotle assembles under the eyes of his reader are all animals: men, quadrupeds, fish, amphibians, birds, insects. He doesn't consider any of these animals either separately or in the classes into which he has arranged them; the whole animal kingdom is for him just a single point: what he writes the history of is *the animal in general* [my emphasis], and if he recounts an observation, particular to one or another animal, it is only either to serve as a proof of a general proposition which he has put forward, or to justify an exception which he warns about. In this way Aristotle, wanting to get to know the nature of animals, first proposes to make an examination of the parts of their bodies, as the first object which strikes the sight: and then, having given some general definitions of these parts, and having distinguished different species among the animals, on the basis of the variety of their exterior forms, he expounds in the first books all the details of the parts of their bodies. The fifth, sixth and seventh books are devoted to explaining in what manner the animal is born; the time when it begins to reproduce, the time when it ceases to be able to reproduce, and the total length of its life. From reading the first seven books, one learns how the animal exists and how it multiplies; the final two books teach how the animal lives and how it conserves itself. The subject of the eighth book is the animal's nourishment, and the places where it lives; the ninth treats its habits (if it is possible to use that expression); Aristotle says there what the habits of the different animals are, with which other animals it lives mutually, whether in peace or in war, and how they are provided for their protection and their defence.
>
> (pp. xii–xiii, my translation)

To Réamur's complaint that Aristotle did not particularise and order his observations better, which would help one remember them, Camus says that describing each animal in detail and distributing them into classes was antithetical to Aristotle's intentions:

> The object of Aristotle was to give *the history of nature in animals*; I say again, he named this or that animal only accidentally and to serve as an example. He mentions many animals because he knew many; but it would have been pointless for his plan to name all of them; his goal was fulfilled when he had justified a general assertion by a certain number of particular facts.
>
> (My emphasis, p. xvii)

It may be a somewhat depressing thought that with every subsequent generation of scholars working on Aristotle and the animal books since the time of
Camus in the 1780s to today, we have got further away from understanding
Aristotle and his animal books, rather than closer. But that seems to be the
case. For the rapid development of new disciplines of life over the last 200
years or so, and the progressive dominance of one after another of the new
disciplines of life as the 'umbrella discipline' of its time, have all put new layers of fog and misunderstanding between us and Aristotle and his animal
books.

Can the translator ever transcend his or her own present in translating
works from the past? Or is the translator doomed always to betray his original, as the Italians claim with their phrase 'traduttore traditore'?

All of this indicates that in interpreting and translating Aristotle, historians and linguists have always had at least one eye on the present – their own
present, that is – and that, with the best of intentions but with no reflection
on what they were doing, they have hauled Aristotle into that present and
made him a modern, a man of their times. But beyond this, as Pierre
Pellegrin has recently shown, historians interested in Aristotle's animal
books have always assumed that there are some 'eternal problems' which
both Aristotle and ourselves (or at least our colleagues in science) have been
trying to solve. But this simply begs the question, for again it is simply an
assumption, not a fact. Pellegrin puts it very well when he says that what we
need to recognise is that Aristotle's work on animals 'is radically foreign to
us: produced in a world that is gone, it tried to answer questions that we no
longer ask' (p. 2).

A recent work on Aristotle's animal books boxes the compass on this issue:
Stéphane Schmitt's *Aux origines de la biologie moderne: L'anatomie comparée
d'Aristote à la théorie de l'évolution* (2006). The title promises all sorts of
confusions. In the first place the origin of modern biology and comparative
anatomy are made one and the same. In the second place Aristotle is credited
with originating both of them. As I hope will now be evident, Aristotle is not
– was not – performing comparative anatomy and was not a comparative
anatomist. In the third place in this book-title comparative anatomy is made
the bridge between Aristotle's supposed 'comparative anatomy' and the modern Darwinian theory of evolution, thus positing a continuity of a problematic and of the practice of a discipline for well over two thousand years. It
would take a whole book to unpick the problems with this title. And yet in
practice, in the text of the book, Schmitt places the origin of comparative
anatomy and biology firmly (and correctly) at the end of the 18th and beginning of the 19th centuries, and his Aristotle is claimed as the originator of
both only in the most general (and one might say, most generous) of terms,
viz. as someone concerned with classifying animals according to their form (a
claim which, unfortunately for Schmitt, happens also to be no longer tena-
ble). 'Aristotle', Schmitt writes, 'appears as the founder of a biology which,
while it is very different from biology today, and is built on philosophical
presuppositions which are largely foreign to us today, evidences a care for

rational explanation which characterises a true science of nature' (p. 17). So the biology of Aristotle is not our biology after all! As Schmitt writes, again,

> it is only in the first years of the 19th century that the term [viz. 'biology'] was employed in the sense of 'science of life', and has little by little emerged in its present-day meaning. This moreover is not insignificant, because it is precisely this period when the vitalist movements developed, which stressed the specificity of life and considered that it was not reducible to physico-chemical laws.
>
> (p. 24)

I should add that once Schmitt reaches what he himself shows is the *actual* period of the origin of comparative anatomy and biology (viz. the very late 18th and early 19th centuries), then his book is easily the best work on the subjects that I know. But why did Aristotle have to get caught up in all this? And why did the practice of biology have to be attributed to him – only for it to be de-attributed later?

Aristotle still today gets represented as a biologist or protobiologist in popular works, in a manner following the kind of analysis offered by D'Arcy Thompson. And in the scholarly community calling Aristotle a biologist has become virtually second nature for scholars studying the animal books. Yet we do not even employ a precise characterisation of biology today in our Aristotle studies which could serve as a point of comparison or assessment with respect to what Aristotle was doing. In using the term 'biology' so freely for Aristotle's project in the animal books, we do not even ask a question such as 'In what ways did Aristotle's enterprise differ from our practice of biology, and in what ways might it be said to resemble biology?' which might yield some interesting information. We don't even open this question when we call Aristotle a biologist. In fact we open no questions at all. Rather, we use the term in a knee-jerk sort of way: if these books are about animals, then they are obviously instances of biology, and Aristotle was therefore a biologist when engaged in the work reported in them.[6] Our tacit working definition of biology is in fact so loose as to be of no practical use: it gives us no key to learning more about these books as instances of biology in practice – if that is indeed what they were.

But are there in fact any *advantages* to using the terms 'biology' and 'biological' with respect to Aristotle's work as represented in the animal books? How does it help us? How does it help Aristotle? The only advantages I can see are (i) we are using a familiar term, whose basic meaning we all roughly understand, so it doesn't need to be explained, and (ii) since that term is the name of a modern scientific discipline, Aristotle thereby gets to sound like a modern, someone like us.

The *disadvantages* seem to me to greatly outweigh these apparent advantages, and to stem directly from them. They are the disadvantages of misrepresentation and of anachronism, and whilst both might help us by making Aristotle sound like ourselves, neither of them helps Aristotle sound like himself. As I have already argued, to label these aspects of Aristotle's work as biology not only begs the question of the nature of his enterprise in the animal books, but it also (whether this is intended or not) draws Aristotle's work into the 20th and 21st centuries and allies it to the modern practice of biology, a practice and a discipline not available to him, and in these ways thus inevitably *misrepresents* what he was up to. I know that some scholars believe that we can still call Aristotle's enterprise in the animal books 'biology', and evade this problem of misrepresentation, for (they claim) we are doing no more than saying that these books deal with 'life' (Greek, *bios*) and its study (Greek, *logos*). But such a claim that our personal use of the term 'biology' with respect to Aristotle's animal books is quite distinct from the use of the term for a modern scientific discipline, is mere naive solipsism. Words such as 'biology' and 'biological' cannot just be given any private meaning we, as individuals, choose. However much we may think we are using the term 'biology' in an innocent way with respect to Aristotle, we cannot avoid being heard to be using it with all the weight of the full-blown scientific discipline that this term was coined to characterise. Underneath all our uses of such terms as biology and biological lies the modern discipline of biology, and the authority of the modern practitioner of biology.

But the issue goes deeper than this, for I think the real problem here is that present-day historians of the animal books positively *want* Aristotle to have been doing biology or some early version of it: and not only does this make him a 'father' of the modern discipline of biology, but the ascription by such historians to Aristotle of the practice of the extremely important modern discipline of biology in its turn is taken to make their man – *Aristotle* – more important historically.[7]

With respect to the anachronism inherent in using the term 'biology' for the enterprise evidenced by Aristotle's animal books, it might seem that we can avoid this by claiming that we are not referring to the enterprise, but only to the individual pieces of information that we might find in those books. Surely we can claim that a simple statement or piece of information is 'biological', without implying or being taken to imply that Aristotle acquired it through the practice of the modern discipline of biology? Surely we are just saying it is a statement about a phenomenon of life (*bios*- and *logos*, again)? Surely all we are offering is simply a description of the status or nature of the information in the modern world? To which my answer would be that we have already committed the anachronism by offering a description of the status or nature of the information *in the modern world*. For one thing we can be certain of is that Aristotle did not function in our modern world, so we are transferring this piece of information from one world to another and inevitably transforming it in the process.

The other side of this issue is that if we go into our studies of Aristotle's animal books without inspecting the disciplinary, activity and intentional terms we are ascribing to him, then we will believe that everything we find confirms our assumption. It would be nice to think that extended immersion in the historical documents would somehow spontaneously lead us to appreciate and recognise the nature of the activities (especially the intellectual activities) that past people were engaged in, but experience does not show many examples of this happening. If we go into our studies thinking Aristotle was practising biology, we're likely to come out thinking the same, however confused or inadequate a biologist we might now think him to have been.

But there is an even bigger problem which arises from our spontaneous knee-jerk classifying of the animal books as works on biology. For once we have separated these books from Aristotle's other books, we become obsessed with questions about the *relationship* of his 'biology' to his 'philosophy'. As we saw, it was a question which struck D'Arcy Thompson after he had separated these books from the others. We agonise over questions such as 'Did Aristotle's biology influence his philosophy, and if so, how?' or its reverse. And when we have finished *contrasting* or *opposing* the supposed biology to the philosophy (and vice versa), we sometimes move on to trying to *reconcile* the two, usually coming to the unremarkable conclusion that they are related. Questions like these have dominated the scholarly literature on Aristotle's animal books in recent years.[8] All these questions arise from *our* action in separating the animal books from the rest. Yet these questions assume that it was *Aristotle* who made this separation. We seek to reconcile in Aristotle's mind and practice things which may never have been separated there.

What we need to do first is ask an open, not a closed, question about what Aristotle was doing, in his own terms. To start our investigations by calling this activity 'biology' is to close the question at the very moment it needs to be opened.

The case is the same with those other apparently innocuous terms, embryology, (experimental) physiology, zoology. As disciplinary labels they, too, are all creations of the very late 18th and early 19th centuries, and were coined in order to characterise new domains of study which were originated in that period. The use of any of them with respect to Aristotle turns him, whether we like it or not, whether we intend it or not, into a modern, indeed into a modern scientist. But he was not a modern but an ancient, not a scientist but a philosopher. He could not practise modern disciplines, only ancient ones.

It may be that recent scholars of Aristotle's animal books will believe that my position differs from their own only in form of words, and will maintain that they, too, of course view the animal books part of Aristotle's larger philosophical project, but that for convenience they choose to refer to this part of that larger project as 'Aristotle's biology' – as a kind of shorthand or place-marker. Are we only arguing over words? I certainly think that some of these modern scholars have done excellent work in recognising and laying out some of the details of Aristotle's goals and procedures in the animal books, even though they have done so whilst insisting on regarding Aristotle as engaged in biology. For instance, Allan Gotthelf and James Lennox, publishing in 1987, might reasonably think that they have said all this themselves, as in the Introduction to the section 'Biology and Philosophy: An Overview' of a book they jointly edited, where they write:

But 'biological' is *our* label: Aristotle has no such term, and speaks rather of the general study of nature (*phusike*), and within that, of the study of plants, or of animals, or of the capacities of soul. Nor must we assume that we can straightforwardly map these studies as Aristotle conceived them onto *portions* of our own general biology, or botany, or zoology (or embryology, or comparative anatomy, etc.) – or even assume that they are *science* rather than philosophy (or philosophy *rather than* science). We need instead to approach them fresh, on their own terms to come to see what their aims are, and their methods, and their contents, and what their relation is to (and how they might be of use in the understanding of) Aristotelian philosophy itself.

(pp. 5–6)

However, even in this laudable attempt not to be anachronistic, the last part of the last sentence reveals that Gotthelf and Lennox are still *contrasting* the animal books to 'Aristotelian philosophy itself', and this caveat paragraph appears in a volume which they have chosen to call *Philosophical Issues in Aristotle's Biology*! That title could hardly be more explicit about their view that Aristotle has a discipline category of 'biology', which one would have thought should have been the very thing at issue rather than something to be assumed in advance. And the first essay in the volume, by the eminent scholar of Aristotle's animal books, David Balme, is called 'The Place of Biology in Aristotle's Philosophy', and makes no mention of soul.

Similarly Christopher Cosans, writing on 'Aristotle's anatomical philosophy of nature' in 1998, too, might think that my point here is adequately made by the title of his own article. But valuable as the points are that he makes, nevertheless in the abstract of his article he claims that 'this consideration of Aristotle as a sophisticated *biologist* helps our reading of his writings in other areas of philosophy' (my emphasis) – that is to say, the work recorded in the animal books is still being seen as separate from and contrasted to the philosophy proper – and the opening words of his article are those now-familiar ones: 'The biological texts of Aristotle …'. And whilst, on the positive side, he prefers to look at the anatomical angle, Cosans nevertheless allows Aristotelian scholars the free use of 'biology' as a term suitable to describe Aristotle's enterprise: 'Although "biology" is a modern word, it has such wide scope that it applies to any scientific investigation of living things' (p. 312). Again, this claim should be just the point at issue, in my opinion, not the casual starting-point of discussion, as if it has already been firmly established.

So, yes, I do think the words matter, especially ones which describe intentional activities and modern disciplines, and I think that if we choose the wrong words we not only make life quite unnecessarily difficult for ourselves, but we end up turning Ancients into Moderns, which is bizarre.

## Notes

1 Thompson, *On Aristotle as a Biologist* (1913).
2 Other scholars tend to place the animal books at the end of Aristotle's life; see for instance Jaeger, *Aristotle*.
3 Gotthelf, 'Darwin on Aristotle', p. 9. Portraying Darwin as both Aristotle and Democritus fits with Ogle's presentation of Aristotle in the Introduction as having to decide between the two (supposedly ancient) positions. Gotthelf discusses this at pp. 22–23.
4 Cuvier is similarly credited as being a comparative anatomist like Aristotle, by William Lawrence in 1816: 'The noble patronage of his pupil Alexander enabled the philosopher to expend an enormous sum in drawing together from all quarters the animals described in this immortal work [viz. *The history of animals*]. He not only knew and dissected a great number of species, but he studied and described them on a vast and luminous plan, to which none of his successors has approached, ranging the facts, not according to the species, but to the organs and functions – the only means of arriving at comparative results. The modern works of Blumenbach and Cuvier, are constructed on the same principle, which was also followed by Mr [John] Hunter in the arrangement of his collection', *An Introduction to Comparative Anatomy and Physiology*, 1816, pp. 32–33.
5 Buffon *Histoire naturelle*, 1749 ed., vol. 1, pp. 49–54 where Buffon credits Aristotle with being one of the first Naturalists, practising Natural History, he does not refer to soul, any more than he does in his own description of what a modern (Buffonian) natural historian does, pp. 34–36.
6 We make a similar knee-jerk identification with respect to the sciences of the organic which come under biology, when we find Aristotle saying things which to us seem to be obviously like those modern-day sciences: physiology, embryology, comparative anatomy, and the like.
7 On the pleasures of promoting 'my man' in writing history, see Cannon, *Science in Culture* (1978), p. 222.
8 See for instance Gotthelf and Lennox, eds., *Philosophical Issues in Aristotle's Biology* (1987); Kullman, *Die Teleologie in der aristotelischen Biologie*, (1979); Kullman and öllinger, eds., *Aristotelische Biologie* (1997).

# 10 Harvey and his historians

## Why and how did we lose this Harvey?

In this chapter I try to open the question of how we lost the understanding of Harvey the investigator of "the animal" and the "vegetative soul", and willing follower of Aristotle, as portrayed by me earlier in the present book.

Harvey's reputation, like those of so many past people whose achievements are still lauded, is a contested one, and the primary reason for studying and writing about Harvey in the first place, even today, for almost anyone, is to make some particular point in the debate. That point and that debate will invariably be about the time in which the historian – not the historical character under study – him or herself is living, and often the point will be about some issue controversial in the day of the historian. In particular in Harvey's case the discussion in the 20th century was about scientific discovery and how it happens and happened, and the relative contributions of genius, on the one hand, and the application of the (supposed) scientific method and measurement, on the other.

Harvey himself, looking back on events between 1628 (the publication of *De motu cordis*) and 1649 (his *Exercitationes* to Riolan), recorded how the discovery was controversial from the very first day:

> Most learned *Riolan*, by the help of the Presse, many years ago, I published a part of my labour: But since the birth-day of the Circulation of the Blood, almost no day has past, nor the least space of time, in which I have not heard both good and evil of the Circulation of the Blood which I found out: Others rail at it, as a tender babie unworthy to come to light; Others say, that its worthy to be foster'd, and favour my writings, and defend them; Some with great disdain oppose them; Some with mighty applause protect them; Others say, that I have abundantly by many experiments, observations, and ocular testimony, confirm'd the Circulation of the blood, against all strength and force of arguments; Others think it not yet sufficiently illustrated, and vindicated from objections.
>
> (*To Riolan 1,* pp. 30–1)

However, according to John Aubrey, Harvey's new doctrine was relatively soon accepted everywhere, 'and, as Mr Hobbes sayes in his book *de Corpore,*

DOI: 10.4324/9781003247616-12

he [viz. Harvey] is the only man, perhaps, that ever lived to see his owne Doctrine established in his life-time'.[1] A number of Harveian historians have traced these disputes over the acceptance or rejection of the doctrine of the circulation, and these do not concern us further here.[2]

There are often three stages in the acceptance of a new doctrine, in whatever subject, both in Harvey's day and in our own. In the first place people say it is mad. In the second they say we have known it all along – and anyway someone else discovered or proposed it. And in the third and last stage people so take it for granted that they can't understand what all the fuss had been about: it was obvious and hardly merits the claim that it was discovered. Harvey's discovery, too, went through these three stages.

The first we have already mentioned, with respect to the attacks (and defences) of the doctrine in Harvey's own lifetime: Harvey must be mad to suggest that all the Ancients, and the tradition of hundreds of years, are wrong – and all the experiments Harvey conducted are wrong or misleading. Aubrey recorded that

> I have heard him say that after his Booke of the Circulation of the Blood came out, that he fell mightly in his Practize and that 'twas believed by the vulgar that he was crack-brained; and all the Physitians were against his Opinion and envied him; many wrote against him, as Dr Primrose, Paracisanus etc.[3]

The second wave of attacks against Harvey's discovery – we have known it all along, and anyway someone else discovered or proposed it – did not begin until after he was dead, and in one form or another has continued to the present. For some people it was obvious that Hippocrates knew it all along. This line of attack is often stimulated by the conscious or unconscious promotion of national heroes to a major role in the business, with Italian scholars promoting claims on behalf of a long-dead Italian physician who lived before Harvey, and Arab scholars similarly promoting claims on behalf of a long-dead Arab physician who also lived before Harvey, and so on.[4]

So: what questions have historians actually raised about how and why William Harvey discovered the circulation of the blood in animals?

It seems to me that there have been three main traditions of looking at Harvey's discovery about the circulation of the blood. I would like to be able to say that this rather blunt characterisation unduly over-simplifies a series of nuanced and sophisticated interpretations, but it doesn't, because only a few of them are particularly nuanced or sophisticated, and those are most recent.

### Jigsaw puzzles and the human spirit

The first of these interpretative traditions is to see the discovery of the circulation of the blood as an eternal problem which just needed someone to put the pieces together to reach the correct conclusion. In this tradition no great store is usually put on Harvey's personal qualities or his supposed genius: he was just the chap who (as it were) put the last piece into the jigsaw. The earliest people who wrote in this way were actually practitioners, one in anatomy, another in experimental physiology. Practitioners of the sciences do not in fact have a privileged view of discovery, or even of regular scientific practice, most of them turning out, when questioned, to have at best simplistic or naive views on these matters. But then their role is not that of historian or philosopher or anthropologist, and if their education as practitioners of the sciences has been a success, it means that they are not much given to introspection, either about their own scientific work or that of others.

The first of these practitioners talking about Harvey was William Hunter, in his *Two Introductory Lectures ... to His Last Course of Anatomical Lectures* in 1783, where he outlined the history of anatomy as he saw it leading up to himself and his contemporaries. Harvey's discovery of the circulation of the blood, says Hunter,

> was by far the most important step that has been made, in the knowledge of animal bodies, in any age. It not only reflected useful lights upon what had already been found out in Anatomy, but also pointed out the means of further investigation.
>
> (p. 42)

But nevertheless, Hunter says, Harvey hardly merits being called a genius in the way that Leonardo da Vinci does:

> For, the singular structure of the parts concerned in the circulation, to wit, the heart, arteries, and veins and the obvious phenomena in bleeding animals to death, the different effects of ligatures on different vessels, the practice of surgery, with regard to bleedings and blood-vessels, the action of the heart when exposed to view in living bodies, all these, I say, so evidently proclaim the circulation, that there seems to have been nothing more required for the discovery, than laying aside some gross prejudices, and considering fairly some obvious truths.
>
> (p. 43)

This interpretation might seem surprising from Hunter – someone who was himself desperate to achieve matching immortality from an anatomical discovery in the way that Harvey had done:

> I think I have proved, that the lymphatic vessels are the absorbing vessels, all over the body; that they are the same as the lacteals; and that these altogether, with the thoracic duct, constitute one great and general system, dispersed through the whole body for absorption ... This discovery gains credit daily, both at home and abroad, to such a degree, that I believe we may now say, that it is almost universally adopted: and, if we mistake not, in a proper time, it will be allowed to be the greatest

discovery, both in physiology and in pathology, that Anatomy has suggested, since the discovery of the circulation.

(pp. 58–9)

But it is clear from Hunter's description of the stream of anatomical discoveries that followed Harvey's discovery, that he did indeed think of discovery in anatomical investigation as simply a matter of building one item upon another, and that "projects of enquiry" (as I call them) did not enter into his considerations about how discovery takes place.

The other practitioner who made early significant comments on Harvey's discovery was Pierre Flourens (1794–1867), who took a different explanatory tack in his 1854 *Histoire de la découverte de la circulation du sang*. Flourens trained in medicine and then had a successful early career as an experimental physiologist in Paris, investigating the phenomena of life by vivisecting the live animal, in the manner that had been recently invented by François Magendie and Claude Bernard.[5] Later in life Flourens appointed himself a historian of the life and physical sciences. He wrote an important biography of Cuvier (one of his own masters), and here in 1854 he published the life of Harvey – or, rather, not of Harvey, but of the discovery itself of the circulation of the blood.

According to his own account, Flourens had been reading a piece by Bernadino Ramazzini, the 18th-century Italian investigator of diseases, which attributed the discovery not to an Englishman but to two Italians. Ramazzini said:

> The ancients were completely ignorant of the circulation of the blood, and we are indebted to Harvey, the English Democritus, for having been the first to publish it, after he had found it in those two excellent sources, Fabricius d'Aquapendent and Paul Sarpi, both of them professors at Padua, and who had performed many experiments about it on all sorts of animals.

At that moment, Flourens recalled, he realised that 'the history of the discovery of the circulation of the blood was still waiting to be done' (pp. 5–6).

For Flourens, that history is a story of errors being put forward, and then being removed: the circulation of the blood is something self-evident, before our eyes, just as soon as these errors have been removed.

> The discovery of the circulation of the blood does not belong, and could not belong, to one man, nor even to a single age. It was necessary to destroy several errors: it was necessary to substitute a truth for each of these errors. All this took place successively, slowly, little by little. Galen first combated Erasistratus; he opened the route which, followed by Vesalius, by Servetus, by Colombus, by Cesalpino, by Fabricius d'Aquapendente, has led us to Harvey.

(p.13)

Flourens elaborates on these three errors which hid the circulation of the blood from broad sight:

> Three principal errors concealed, if I can put it like that, the great fact of the circulation of the blood: the first, that the arteries contain only air; the second that the septum which separates the two ventricles [of the heart] is pierced; the third, that the veins carry the blood to the parts, rather than bring it back from them. Let us see who are the men who put forward these errors, and who are the men who destroyed them.
>
> (pp. 13–4)

For Flourens, as for many of his French contemporary men of science, it was *l'esprit humain* which was the real star of all this history: 'The history of science is the CHRONOLOGY of the human spirit', as he puts it, using appropriate capitals. (p. 12). In such history 'One sees the succession of the advances, the sequence of the names, the filiation of the ideas' (*'la suite des progrès, l'ordre des noms, la filiation des idées'*) (p. 12).

The human spirit as an active agent appears on many pages of Flourens' histories. For instance, in his 1845 book on the work of Cuvier, he describes Cuvier's own history of the sciences as being 'l'histoire même de l'esprit humain; car, remontant aux causes de leurs progrès et de leurs erreurs, c'est toujours dans les bonnes ou mauvaises routes, suivies par l'esprit humain, qu'il trouve ces causes'. Ingenious systems might come and go, but only the facts matter, discovered through observation and experiment by 'the human spirit' operating through a sequence of human agents.

In brief, this is the succession of ideas, and the individuals that promoted them, that Flourens pithily tells about the discovery of the circulation of the blood, and which is correct as far as it goes:

> Galen, who proved that the arteries contain blood, and not air, as Erasistratus believed; Vesalius, who proved that the septum is solid and not pierced, as Galen believed; Servetus, Colombus, Cesalpinus, who proved that the blood of the heart passes via the lung before returning to the left heart, a passage which constitutes the *pulmonary circulation*; Cesalpino who, first of all, saw that the blood in the veins returns from the parts to the heart, instead of going from the heart to the parts, a return (*retour*) which constitutes the *general circulation*; Fabricius d'Aquapendente who, first of anyone, saw the valves of the veins, without understanding their function; and finally Harvey, a man excellent at demonstrating things first perceived by other people, who proved the *pulmonary circulation* by the very structure of the heart, the *general circulation* by the arrangement of the valves of the veins, who reconnected these two circulations, one to the other, and who gave us the complete view of this great mechanism.
>
> (p. 9, my translation)

But the individuals here are nothing special, merely the agents of the human spirit.

So when he comes to Harvey, what Flourens has to say about him is quite minimal. The most important thing about him was that he went to Padua for his education, where, according to Flourens, 'the state of the question was known to everyone, where everything that had been said about the circulation was known to everyone' (p. 43). C'est tout! Which, to my 21st-century mind, is actually no explanation at all – of anything!

The two most important features of Flourens' account for our present purposes are: first, that he represents the search for the circulation as an eternal question, one with which the human spirit has been struggling since time immemorial. But as we have already learnt, there are no eternal questions (See Introduction in this book). And the second important feature is actually an absence: he does not mention Aristotle and his place in Harvey's thinking.

## Scientific method and experiment

The second dominating theme in histories of Harvey and his discovery, for almost two centuries now, has been an obsession with 'the scientific method', and how Harvey must have been practising it, because that's the only way great discoveries are made, is it not? In the early to mid-19th century it began to be a central concern of philosophers and practitioners, especially in the Anglophone tradition, to establish what it is that distinguishes the sciences of nature and number from other types of human enquiry. First (they argued), there is their subject-matter, second, their objectivity, and third (supposedly), the method of enquiry used in them. So a number of philosopher-historians set out to define the scientific method, and then to trace its history: to find who first enunciated it, who improved it, who successfully deployed it, what it looks like in action, and so on.

As might have been expected, the scientific method proved very elusive and contradictory to these historians and philosophers: it was both inductive and deductive, both experimental and a product of detached reasoning, both X and Y, and agreement has still not been reached on whether there is in fact a single definable scientific method at all. Attempts to find its origin have had equally mixed success. It seemed to John Stuart Mill, William Whewell and John Herschel that whatever this special approach was, it could be traced back to Francis Bacon in the beginning of the 17th century. But subsequent scholars have placed its origin elsewhere – with an earlier Bacon, Roger, in the 13th century, or with professors at Padua in the 16th century, and so on.

The basic account is given by Robert Willis, in his *Life* of Harvey, prefaced to his translation of the *Works*, 1847. Willis' writings on Harvey and his translations of the works into English have made him a mighty influence in interpretations of Harvey. He makes Harvey the discoverer both a Baconian (Francis), a modern experimental physiologist, and also a genius. Harvey succeeds by using 'induction upon data carefully collected and considered;

and it would not be easy to adduce a more striking example of the way in which ultimate truth is arrived at by a succession of inferences than is contained in Harvey's Essay on the Heart and Blood' (p. xl). Willis' assessment of Harvey follows the teachings of his own teacher, John Barclay the Edinburgh anatomist: it was

> by superior intellect, that Harvey made his immortal discovery ... Harvey's discovery was of the rational and inductive and therefore higher class, according to our estimate; it was made in virtue of the intellectual powers which particularly distinguish man, possessed in a state of the highest perfection.

In another biography of Harvey that Willis published in 1878, *William Harvey: A History of the Discovery of the Circulation of the Blood*, Willis considers and dismisses the claims that might be made for Harvey's 'predecessors'. But of Harvey he writes 'He alone of all [students at Padua] was privileged by partial Nature's fiat to put to interest the lessons of his teachers, to divine the goal to which ever accumulating facts were pointing, and through them to conquer immortality for himself' (p. 158).

> Civilized Europe, ancient and modern, had been slowly contributing and accumulating materials for its production [viz. the idea of the circulation of the blood]; Harvey at length appeared, and the idea took fashion in his mind and emerged like Pallas in panoply from the brain of Jove
>
> (p. 161)

It was '[an] "Idea" which I have been tempted to think he brought with him from Padua, still in the germ or but half evolved ...' (pp. 162–3). So Willis manages to box the compass. For him Harvey was a special genius, effectively using the Baconian scientific method, and coming up single-handedly with an idea which solved all the (supposed) problems of other investigators. And as there is no explaining genius, so Willis does not seem to need Harvey to be tackling any specific problem in his anatomising.

By contrast, when we turn to Kenneth D. Keele, and his *William Harvey: The Man, the Physician, and the Scientist* (London, 1965), we do find him seeking to open up such questions. Keele was a physician (MD and FRCP) and in 1929, as a student at St Bartholomew's Hospital (where Harvey had worked) he had received the Harvey Prize in Physiology. Ever since, he sought answers to questions such as 'How did Harvey make his great discovery of the circulation? What were its effects in his own and later times? What is his place in the long history of scientific progress?' Keele believed that the role of Aristotle in Harvey's thinking had been downplayed.

> In order to appreciate Harvey's work on the movement of the heart and blood it is necessary to have a clear idea of *the problem which confronted him* [my emphasis]. To reach this one must briefly trace the story of

previous ideas of the functions of the heart and blood – the prelude to Harvey's great discovery

(p. 109)

But all this just involves a run-through of the traditional usual suspects: Aristotle, Erasistratus, Galen, Ibn al Nafis, Leonardo, Vesalius, Servetus, Colombo, Caesalpino and Fabricius. In the end Keele does not answer his initial question of how Harvey made the discovery, nor does he do much to reinstall Aristotle in the story. Harvey seemingly 'sees himself as a conscientious observer of phenomena applying the inductive form of reasoning given him by Aristotle ... for Harvey scientific investigation consisted exclusively of a series of inductions from observations' (p. 107), which effectively makes him both an Aristotelian and a Baconian!

Geoffrey (later Sir Geoffrey) Keynes (1887–1982) was a doctor, an innovator in blood transfusion and a bibliophile. He was, in effect, the doctor to the members of the Bloomsbury Group (of which his more famous brother Maynard was a key member), so he was regularly in touch with artistic innovations. But his own preferred role in these affairs was to assemble bibliographies and biographies about dead people. His favourite past writers who received these attentions included William Blake, Rupert Brooke (a schoolfriend of his), Jane Austen, William Hazlitt and others, while his favourite past medical men included Sir Thomas Browne, Sir Astley Cooper, Ambroise Paré, Martin Lister and, of course, William Harvey. These scrupulous and totally trustworthy monuments of erudition were assembled by Keynes in the little leisure afforded by a busy medical career. As far as I can judge Keynes's interest in Harvey came from his own involvement in developing blood transfusion (a practice originally, if briefly, inspired by Harvey's discovery), and the fact that he himself held posts at St Bartholomew's Hospital in London, as Harvey himself had done.

Keynes' careful and judicious biography of Harvey (1966) is now our primary resource for information about Harvey's life. But Keynes is completely reticent on the nature and point of Harvey's researches, if he is interested in them at all directly. This is all he says about Harvey taking up research:

His time in Padua and his work under Fabricius had inspired him with a passion for "research" ... Fabricius had taught him the pleasures of discovering new facts by the dissection of human bodies and the intellectual satisfaction of identifying anatomical and biological parallels through the investigation of the bodies of other animals. If his time in these early years was not fully occupied by a growing medical practice, then we may be certain that Harvey was accumulating notes and thinking over the problems ultimately to be crystallized in his first published work on the circulation of the blood.

(p. 49)

To which one can only say: 'accumulating notes about what?', and 'thinking over what problems?' In other words, for Keynes there is no question to be raised here about Harvey's research or why he is doing any, even though Harvey's contemporary John Aubrey – whom Keynes quotes at length elsewhere in his book – points out the uniqueness of Harvey's position in doing anatomical research, saying that Harvey 'was the first that I hear of that was curious in Anatomy in England' (p. 432).

So my conclusion as to why we lost the Harvey who was dedicated to Aristotle and who consequently discovered the circulation of the blood, is that we had already lost this Aristotle, while Harvey's historians were anyway intent on something else in offering their accounts of the discovery.

In terms of investigating and exploring what Harvey was doing and why in the anatomical research that led him to discover the circulation of the blood, there have been several outstanding scholarly contributions from the 1950s onwards, though none of them pose the question of what Harvey was doing in quite the way that I would wish. Aristotle has come to be more significant in Harvey's thinking in these writings.

Walter Pagel was publishing on Harvey and the purpose of the circulation back in 1951 when he attributed Harvey's concern with purpose to his committed Aristotelianism. He continued to study Harvey into the 1970s when he published *New Light on William Harvey* (1976). Pagel was also concerned with Paracelsian and mystical medicine in Harvey's lifetime, and was particularly interested in Harvey's use of the macrocosm/microcosm comparison with respect to the heart and the body, the sun and the world, the king and his kingdom, and with the whole concept of circulation.

Gweneth Whitteridge was very active in the two or more decades from the mid-1950s in transcribing and translating the surviving Harvey manuscripts, as well as producing new English editions of Harvey's main works. She published her biography of Harvey in 1971, *William Harvey and the Circulation of the Blood*. In its scholarship it towers over all previous work on Harvey. But here she is adamant that 'Harvey falls into the category of the great scientist who is not conscious of any philosophical method underlying his actions'. This approach is confirmed for her by Harvey's stress on observation, experiment and trust in the senses. For her this is the role that Aristotle played for Harvey: 'There can be little doubt that Harvey's great admiration for Aristotle rests primarily on Aristotle's capacity for patient observation of Nature' (Preface). Other historians could see this as a rather philosophically naïve approach on her part.

Since 1965 Charles Webster has been producing a rich stream of articles on Harvey, the College of Physicians in his time, the origins of blood transfusion, Harvey's conception of the heart as a pump, and related topics. He has been very concerned to re-establish the social and political context in which investigations of nature were carried out in 17th-century England, and Harvey's role as a model for research by other people.

Most recently Roger French has written on *William Harvey's Natural philosophy* (1994). In Harvey's time natural philosophy was a highly contested area, and the historic role of Aristotle in university education was being challenged on all sides. French locates Harvey's position with respect to other natural philosophers of his time, concluding that Harvey's contribution to a theocentric natural philosophy was nothing less than experiment.

However, recent popular/scholarly accounts do not take us any further and rely mostly on the works we have just been discussing. Andrew Gregory in his *Harvey's Heart: The Discovery of Blood Circulation* of 2001 (Duxford, Cambridge) gives Harvey no problem to solve, though he does raise the question. Aristotle certainly appears in Gregory's book (following the work of Roger French and myself). Gregory continues to ask:

> How, though, did Harvey, arrive at the theory of the circulation? Harvey's best argument for the circulation of the blood, and certainly his best known, is that there is so much blood flowing through the heart that the blood must circulate.
>
> (pp. 54–5)

But that's it. In a series aiming to examine 'the thought process leading to his or her discoveries', Jole Shackelford in his 2003 account *William Harvey and the Mechanics of the Heart* (Oxford, 2003) certainly has Harvey asking questions, yet with no specific focus except that he's following Fabricius. Or rather, Shackelford derives Harvey's focus from the results he gets. Finally Thomas Wright's 2013 book, *William Harvey: A Life in Circulation* (or *Circulation: William Harvey's Revolutionary Idea,* Oxford) is an excellent and lively summary of the state of the field, and he shows Harvey following the 'Aristotle project' (as I first called it), but otherwise reveals nothing hitherto unknown about Harvey's work and goals. But I doubt whether any of these more popular writers would claim to be offering ground-breaking new research.

## Notes

1 As quoted in Keynes, p. 435, from the Aubrey manuscripts in the Bodleian Library.
2 See most recently Roger French, *William Harvey's Natural Philosophy*, 1994.
3 As quoted in Keynes, p. 435, from the Aubrey manuscripts in the Bodleian Library. 'Paracisanus' is Parisanus.
4 In this context the supposed role that Ibn Nafis, the 12th-century polymath working in Egypt, had in discovering the so-called "pulmonary circulation', has happily been recently settled by a young scholar, Nahyan Fancy. He has done this not by giving some final decision about who discovered it first, but by showing that the question is irrelevant and misconceived: if we look properly to see what Ibn Nafis was actually doing, we find that 'for Ibn al-Nafis, the pulmonary transit of blood is no more than an anatomical corollary to his new understanding of the soul-spirit-body relationship and its consequent physiology. It is not, as it is for Harvey and modern medicine, a key ingredient (if not the starting point) for a circulatory physiology' (p. 109).
5 On which see Cunningham, *The Anatomist Anatomis'd*, Chapter 6.

# Appendix

## English or Latin?
A note on the editions and translations of Harvey's published works.

In the *Bibliography of the Writings of William Harvey* that he assembled in 1928 and revised in 1952, Geoffrey Keynes wrote that 'the Latin text [of *De motu cordis*] has been printed twenty-three times, including its appearances in the collected works and in facsimiles of the first edition' (p. 5). These Latin versions of the text were the ones read by Harvey's contemporaries, friend and foe alike, and by other doctors in the next century and a half. In 1766 the Royal College of Physicians published a Latin version, edited by Mark Akenside, which has claims to being the most accurate.

It is understandable therefore why Harvey scholars of the last two centuries have assumed that the 17th-century Latin version was the original, and why they thought that modern audiences, mostly lacking Latin, deserved new English translations from the Latin of the epoch-making book on the motion of the heart and blood in animals. As one might expect, all the translators, like many of the biographers of Harvey, had connections with the teaching of experimental physiology, and most of them also with the Royal College of Physicians and St Bartholomew's Hospital in London, both places with very strong links with William Harvey.

However in the present book I take the 17th-century English editions of Harvey's two great published works, *The Anatomical Exercises … Concerning the Motion of the Heart and Blood* of 1653 and the *Anatomical Exercitations Concerning the Generation of Living Creatures* of 1651, as being the originals as written by Harvey himself, and I quote them accordingly. To my knowledge no manuscript originals for these works, in either English or Latin, survive, so we can only deal with the printed editions. I believe that both of these books were written in English by Harvey, and then both were translated into Latin, with the Latin versions being published first in both cases.

We have no clue as to who translated the book on the motion of the heart and blood into Latin. The only specific account of who translated any of Harvey's works into Latin comes from John Aubrey, Harvey's contemporary. Aubrey wrote,

> *The Circuit. Sang. Was, as I take it, donne into Latin by Sir George Ent (quære) as also his Booke de Generatione Animalium, but a little book in*

> 12° *against Riolani (I thinke), wherein he makes-out his doctrine clearer,*
> *was writ by him selfe, & that, as I take it, at Oxford.*[1]

Let us look at these in sequence.

First, '*The Circuit. Sang. Was, as I take it, donne into Latin by Sir George*
*Ent (quære) ...*'. The dates are quite wrong for this, as Ent was still a student
at Cambridge in 1628 when Harvey's book was published in Latin, and he
seems not to have met Harvey before 1636 or 1637, when they met in Venice.

However, Aubrey's note continues, 'But The Circuit. Sang. Was, as I take
it, donne into Latin by Sir George Ent (quære) *as also his Booke de Generatione*
*Animalium ...*'. This second claim is very likely to be true. In the first place
Ent himself says he received the text from Harvey's own hand, and that he
says he undertook to bring it to the press. However, Ent does not say that the
text when Harvey handed it to him was in English. (By contrast Keynes suggests, though without evidence, that Ent translated the text from Harvey's
Latin into English! See Keynes, *Bibliography* p. 49.)

On the other hand, we know incidentally that Ent translated Francis
Glisson's 1654 book on the anatomy of the liver from English into Latin,
though there is no mention of this in the printed edition, so the lack of any
acknowledgement to Ent by Harvey, or to Ent by himself, is not an adequate
argument from silence.[2]

Aubrey's note continues, '*but a little book in 12° against Riolani (I thinke),*
*wherein he makes-out his doctrine clearer, was writ by him selfe, & that, as I*
*take it, at Oxford*'. This refers to the *Exercitatio Anatomica [or, Exercitationes*
*duae Anatomicae] de Circulatione Sanguinis. Ad Joannem Riolanum filium ...,*
which was published in Latin in 1649 at both Cambridge and Rotterdam.
However, an English version of this was published in London in 1653 as *Two*
*Anatomical Exercitations Concerning The Circulation of the Blood, To John*
*Riolan the Son ...* It would take a better Latinist than me to work out whether
the English version was the original, written by Harvey, or whether the Latin
version was the original, written by Harvey, but I know where my inclination
lies.

So, for me the English is the original, the Latin the translation.[3] This turns
upside down all the assumptions on which the translators of Harvey's writings have worked. The manuscript of the English book, *The Anatomical*
*Exercises ... Concerning the Motion of the Heart and Blood [in Animals]*
(1653, republished in 1673), was turned into *Exercitatio anatomica de motu*
*cordis et sanguinis in animalibus* (1628).

Here are the translations of this book made by scholars from the Latin
into English. If I am right, this means that the original 17th-century English
was translated into 17th-century Latin by an unknown person and then, in
the 19th and 20th centuries, this Latin version was translated into more modern English. In other words, if I am correct, they have all gone from English
to Latin to English, a double translation. All the modern translators of the
text into English seem to be motivated to taking up the task, at least in part,
by a desire to inspire modern medical students in some way.

1. By Michael Ryan, 1832-3, in the reforming journal he edited, *The London Medical and Surgical Journal; exhibiting a view of the improvements and discoveries in the various branches of medical science* vols. 1 & 2. Of this Whitteridge wrote 'This version is of no particular merit' (Whitteridge 1971, p. ix, see below). Ryan says nothing specific about why he has translated the book nor why he is publishing it, though something may perhaps be inferred from the title of the journal.[4]

2. By Robert Willis, 1847.[5] As part of *The Works of William Harvey, M.D.* Made at the request of the Sydenham Society. Reprinted (with or without editorial modifications by others) 1889, 1894, 1907, 1949, 1962, 1965, 1989, 2016, and perhaps further reprintings. In his Preface Willis says that he had originally thought that Harvey wrote in English 'and that the English versions of his writings were the proper originals, the Latin versions the translations'. But when he started his project, Willis came to the conclusion that the English version

   > was extremely rebutting in point of style and full of egregious errors, and that nothing short of an entirely new translation could do justice to this admirable treatise, or secure for it, at the present day, the attention it deserved.

   He does not call attention to any of these errors, egregious or otherwise. He used the 1766 RCP edition of the Latin. In turn this translation has been criticised for 'egregious errors' by Geoffrey Keynes in 1953, though he also does not specify any of these errors.[6] Shortly before this project on Harvey, Willis, a practising doctor, had translated the *Elements of Physiology* from the German of Rudolph Wagner, which is a work on experimental physiology.

3. By Chauncey Leake, 1928, to celebrate the 300th anniversary of the Latin publication of the book. Leake was an American professor who worked primarily as a pharmacologist, with a Ph.D. also in physiology. He writes,

   > this volume has been prepared chiefly in the hope that it may interest medical and advanced zoological students, by offering in a dignified but inexpensive way an opportunity to become acquainted, intellectually, with one of the greatest contributors to their subjects.

   It is, he writes, a 'deliberate attempt to present Harvey's thought in the current physiological manner'. In the Preface to the 3rd edition (1941) Leake says, 'Harvey's great classic stands solidly and large in the foundation of modern science. Not dictated by any economic or political consideration, it challenges recent sweeping claims of dialetic materialists regarding the evolution of science', so he may have had an ideological reason too for making this translation. The 5th edition (1970) includes the prefaces to all the previous editions.

4.  By Kenneth J. Franklin, 1957. Franklin was Professor of [Experimental] Physiology at St Bartholomew's Hospital. Produced to coincide with the celebration of the tercentenary of Harvey's death. He writes that

> earlier versions were not sufficiently accurate for those doing either historical or experimental research upon the circulation ... I imagined it would probably be worth while to get nearer to him in thought by attempting to discover more precisely what he intended to convey by every Latin sentence that he included in his great work of 1628. My expectation has been justified ....

He says, 'I have used the standard 1766 edition of the Latin text', and in this first edition by Franklin that Latin text is included.[7] This version has subsequently appeared under different titles. In 1957 as *Movement of the Heart and Blood in Animals. An Anatomical Essay by William Harvey. Translated from the original Latin by Kenneth J. Franklin and now published for the Royal College of Physicians of London,* Oxford. Then in 1963, as *William Harvey: The Circulation of the Blood, and Other Writings,* London and New York, Everyman's Library edition – including the two letters to Riolan. A 1990 edition of it has an introduction by Andrew Wear.

5.  By Gwenneth Whitteridge, 1976, as *William Harvey: An Anatomical Disputation Concerning the Movement of the Heart and Blood in Living Creatures* (Oxford, 1976), as part of her large project on the manuscripts and translations of Harvey's works. Her main concern in producing this version seems to have been to establish that the book was written and should be read as an academic disputation, an argument against the interpretations of certain other modern scholars. She says it is based on the 1653 'anonymous translation' into English, and that 'I have not made use of any modern technical phraseology in matters anatomical or physiological'. She was married to a Professor of (Experimental) Physiology who worked successively at Edinburgh and then Oxford Universities which suggests that she would read Harvey as a practitioner of an early version of experimental physiology.

## Letters of Harvey

Six Letters from Harvey, printed in Latin in the *Opera Omnia* of 1766, have been translated by K. J. Franklin and published in *William Harvey: The Circulation of the Blood,* London, 1963. Charles Webster, in his entry on Harvey in *The Dictionary of Scientific Biography* says that the editors of the 1766 *Opera* had access to Ent's transcription of Harvey's letters. Their present whereabouts does not seem to be known.

## Notes

1  As transcribed from the Aubrey manuscripts in the Bodleian Library in Oxford, and printed as Appendix I in Keynes, *The Life of William Harvey* (1978), p. 433.
2  Cunningham, *English Manuscripts of Francis Glisson* (1993), pp. 3–4. Glisson gave these lectures in English. 'Truly I had rather begun the business anew, than to have been tied to translate what I had done before ... my worthy friend Dr Ent being present, he freely offered his assistance in it ... I with much thankfulness accepted of his kind offer ... All which my worthy friend Dr Ent, with a great deal of patience and resolution, taught to speak Latin'. In the final, Latin, edition of Glisson's book, Ent's role is not mentioned.
3  Here I am in agreement with the arguments to this effect put forward by White in 'The 1653 English edition of *De motu cordis*' (1999), pp. 65–91.
4  On Ryan see Desmond, *The Politics of Evolution* (1989).
5  On whom see Hildebrand, *Robert Willis (1799–1878): The works of William Harvey, M. D.* (2007).
6  Editor's Postscript to Keynes's edition of *The Anatomical Exercises*, p. 199 in the 1995 reprint.
7  Franklin gives an extended account of his experience in Franklin, 'On translating Harvey' (1957).

# Select Bibliography

Adelmann, Howard B., *The Embryological Treatises of Hieronymus Fabricius of Aquapendente*, 2 vols., Cornell, 1942, republished 1967.

Aristotle, *The Works of Aristotle translated into English under the Editorship of J.A. Smith [and] W.D. Ross. Volume IV Historia Animalium, by D'Arcy Wentworth Thompson*, Oxford, 1910.

Aristotle, *History of Animals*, tr. A.L. Peck, Loeb edition, 1970.

Aristotle, *De Anima* (On the soul), tr. H.S. Hett, Loeb edition 1936, repr. 1975.

Aristotle, *Parts of Animals*, tr. A.L. Peck, Loeb edition 1937, repr. 1983.

Balme, D. M., 'Aristotle's use of differentiae in zoology' (1962), as reprinted in Barnes, J., Schofield, M., and Sorabji, R. (Eds.), *Articles on Aristotle 1. Science*, London, 1975, pp. 183–193.

Beyer, Johannes Hartmann, *Pentateuchos Cheirurgicum Hieronymi Fabricii ab Aquapendente*, Frankfurt, 1604.

Boyle, Robert, *Memoirs for The Natural History of Humane Blood, especially the Spirit of that Liquor*, London, 1683/4.

Caius, *An Autobibliography by John Caius*, V. Nutton (Ed.), Abingdon, 2018.

Cannon, Susan Faye, *Science in Culture. The Early Victorian Period*, New York, 1978.

Casserius, Julius, *De Vocis Auditusque Organis Historia Anatomica*, Ferrara, 1600–1.

Charleton, Walter, *The Immortality of the Human Soul, Demonstrated by the Light of Nature*, London, 1657.

Cherniss, Harold, *Aristotle's Criticism of Presocratic Philosophy*, Baltimore, 1933.

Comte de Buffon, George Louis Leclerc, *Histoire Naturelle, Générale et Particulière, avec la Description du Cabinet du Roi*, Paris, 1749.

Corcilius, Klaus, 'Soul, parts of the soul, and the definition of the vegetative capacity in Aristotle's *De anima*', in *Vegetative Powers*, Baldassarri, F. and Blank, A. (Eds.), Cham, Switzerland, 2021, pp. 13–34.

Cornford, Francis, *Before and After Socrates*, orig. publ. 1932, Cambridge, 1979.

Coryat, Thomas, *Coryat's Crudities Hastily Gobbled up in Five Moneths Travells in France*, etc, 1611, repr. Glasgow, 1905.

Cosens, Christopher, 'Aristotle's anatomical philosophy of nature', *Biology and Philosophy*, 13, 1998, 311–339.

Cunningham, Andrew, 'Fabricius and the "Aristotle Project" in Anatomical Teaching and Research at Padua', in *The Medical Renaissance of the Sixteenth Century*, Wear, A., French, R.K., and Lonie, I.M. (Eds.), Cambridge, 1985, pp. 195–222.

Cunningham, Andrew, *The Anatomical Renaissance: The Resurrection of the Anatomical Projects of the Ancients*, Aldershot, 1997.

Cunningham, Andrew, 'The pen and the sword: Recovering the disciplinary identity of physiology and anatomy before 1800. Part I: Old physiology: The pen', *Studies in History and Philosophy of Science*, 33, 2002, 631–665.

Cunningham, Andrew, *The Anatomist Anatomis'd: An experimental discipline in Enlightenment Europe*, Farnham, 2010.

Cunningham, Andrew, 'The principality of the blood: William Harvey, the blood, and the early transfusion experiments', in *Blood – Symbol – Liquid*, Santing, Catrien and Touber, Jetze (Eds.), Leuven, 2012, pp. 193–205.

De Angelis, Simone, 'From text to the body. Commentaries on *De Anima*, anatomical practice and authority around 1600', in *Scholarly Knowledge: Textbooks in Early Modern Europe*, Campi, Emidio, De Angelis, Simone, Goeing Anja-Silvia, and Grafton, Anthony T. (Eds.), 2008, pp. 205–225.

Debus, Allen G., *The Chemical Philosophy: Paracelsian Science and Medicine in the Sixteenth and Seventeenth Centuries*, 2 vols., New York, 1977.

Descartes, *Philosophical Letters*, tr. Anthony Kenny, Oxford, 1970.

Desmond, Adrian, *The Politics of Evolution: Morphology, Medicine and Reform in Radical London*, Chicago, 1989.

Dewhurst, Kenneth, *Richard Lower's* Vindicatio [of Willis's *Diatribae duae*]. *A Defence of the Experimental Method*, Oxford, 1983.

Durrant, Michael, Ed., *Aristotle's De Anima in Focus*, London, 1993.

Fabricius, Hieronymus, *De Visione, De Voce, De Auditu*, Venice, 1600. Part of his *Theatrum totius Animalis Fabricae*.

Fabricius, Hieronymus, *De Venarum Ostiolis*, see Franklin.

Fabricius, Hieronymus ab Aquapendente, *Pentateuchos Cheirurgicum*, Johannes Hartmann Beyer (Ed.), Frankfurt, 1604.

Fabricius, Hieronymus ab Aquapendente, *Opera Chirurgica*, Venice, 1619.

Fancy, Nahan, *Science and Religion in Mamluk Egypt: Ibn al-Nafis, Pulmonary Transit and Bodily Resurrection*, London, 2013.

Favaro, A., *Atti della Nazione Germanica Artista nello Studio di Padova*, 2 vols., Venice, 1911.

Fernel, Jean, *Medicina, Physiologiam, Methodumque Complectens*, Venice, 1555.

Flourens, Pierre, *Histoire de la Découverte de la Circulation du Sang*, Paris, 1857 (1854).

Frank, Robert G., Jr., *Harvey and the Oxford Physiologists: A Study of Scientific Ideas*, London, 1980.

Franklin, K. J., *De Venarum Ostiolis 1603 of Hieronymus Fabricius of Aquapendente*, Springfield, Ill., Thomas, 1933.

Franklin, K.J., 'On translating Harvey', *Journal for the History of Medicine*, 12, 1957, pp. 114–119·

French, Roger, *The History of the Heart: Thoracic Physiology from Ancient to Modern Times*, Aberdeen, 1979.

French, Roger, *William Harvey's Natural Philosophy*, Cambridge, 1994.

Fuchs, Thomas, *The Mechanization of the Heart: Harvey and Descartes*, translated from the German by Marjorie Grene, Rochester, 2001.

Gasking, Elizabeth, *Investigations into Generation, 1651–1828*, London, 1967.

Gilson, Etienne, *History of Christian Philosophy in the Middle Ages*, London, 1955.

Glisson, Francis, *English manuscripts of Francis Glisson (1): From Anatomia Hepatis (The Anatomy of the Liver)*, 1654, Cunningham, Andrew (Ed.), Cambridge, 1993.

Goodall, Charles, *The Royal College of Physicians of London*, London, 1684.

Gotthelf, Allan, Ed., *Aristotle on Nature and Living Things: Philosophical and Historical Studies Presented to David M. Balme on his Seventieth Birthday*, Pittsburgh, 1985.

Gotthelf, Allan, 'Darwin on Aristotle', *Journal for the History of Biology*, 32, 1999, 3–30.

Gotthelf, Allan and Lennox, James G., Eds., *Philosophical Issues in Aristotle's Biology*, Cambridge, 1987.

Grene, Marjorie, *A Portrait of Aristotle*, London, 1963.

Harvey, *William Harvey: An anatomical disputation concerning the movement of the heart and blood in living creatures. Translated with Introduction and notes by* Gweneth Whitteridge, Oxford, 1976.

Harvey, William, *The Anatomical Exercises of Dr William Harvey Professor of Physick, and Physician to the Kings Majesty, Concerning the Motion of the Heart, and Blood, in Living Creatures* (see p. 1 for this title), London, 1653a. First published in Latin as *Exercitatio anatomica de motu cordis et sanguinis in animalibus, Guilielmi Harvei Angli Medici Regii, & Professoris Anatomiae in Collegio Medicorum Londinensi*, Frankfurt, 1628. Here referenced as **Motion**.

Harvey *To Riolan 1 & 2* = Harvey, William, *Two anatomical exercitations concerning The Circulation of the Blood*, To *John Riolan* the Son, the most experienced Physician in the Universitie of *Paris*, the Prince of Dissectors of Bodies, and the Kings Professor and Dean of Anatomie, and the knowledge of Simples; Chief Physician to the Queen-Mother of *Lewis* XIII. *The Author*, William *Harvey*, an *Englishman*, Professor of Anatomie and Chirurgerie in the College of Physicians at London, and Doctor of Physick to the Kings most Excellent Majestie, London, 1653b. Here referenced as **To Riolan 1, 2.**

Harvey, William, *Praelectiones Anatomiæ Universalis*, Edited with an autotype reproduction of the original, by a Committee of The Royal College of Physicians of London, London, 1886. Here referenced as **Lectures** or **Praelectiones**. See also Whitteridge, below.

Harvey, William, The Third Letter. In Reply to Robert Morison, M.D. of Paris, 1653, tr. Kenneth J. Franklin, in *William Harvey: The Circulation of the Blood*, London, 1963, pp. 193–200.

Harvey, William, *Two Anatomical Exercitations Concerning The Circulation of the Blood*, To *John Riolan*, London, 1973, see *To Riolan*, Two Letters.

Harvey, William, *Anatomical Exercitations, Concerning the Generation of Living Creatures*: To Which is added Particular Discourses, of *Births*, and of *Conceptions* etc, London, 1653. First published in Latin as *Exercitationes de generatione animalium. Quibus accedunt quaedam de partu: de membranis ac humoribus uteri: & de conceptione, Autore Guilielmo Harvei Anglo*, London, 1651. Here referenced as **Generation**.

Hildebrand, Reinhard, 'Robert Willis (1799–1878): *The works of William Harvey, M. D.*, London 1847 A bibliographical note', *Sudhoffs Archiv*, 91, 2007 118–121.

Hornblower, Simon and Spawforth, Anthony, (Eds.) *The Oxford Classical Dictionary*, Oxford, 1996.

Hunter, William, *Two Introductory Lectures, Delivered by Dr. William Hunter, to his Last course of Anatomical Lectures*, London, 1784.

Jaeger, Werner, *Aristotle; Fundamentals of the History of his Development*, Tr. Richard Robinson, Oxford, 1934.

Jardine, N., 'Galileo's road to truth and the demonstrative regress', *Studies in History and Philosophy of Science*, 7, 1976, 277–318.

Keynes, Geoffrey, *A Bibliography of the Writings of Dr William Harvey 1578–1657*, Cambridge, 1928, repr. 1953.

Keynes, Geoffrey, *The Life of William Harvey*, Oxford, 1978, first published 1966.

Kullman, Wolfgang, *Die Teleologie in der aristotelischen Biologie: Aristoteles als Zoologe, Embryologe und Genetiker*, Heidelberg, 1979.

Kullman, Wolfgang and Föllinger, Sabine, Eds., *Aristotelische Biologie: Intentionen, Methoden, Ergebnisse*, Stuttgart, 1997.

Lawrence, William, *An Introduction to Comparative Anatomy and Physiology*, London, 1816.

Lee, Robert, *History of the Discoveries of the Circulation of the Blood, of the Ganglia and Nerves and of the Action of the Heart*, London, 1865.

Lewes, George Henry, *Aristotle: A Chapter from the History of Science, including Analyses of Aristotle's Scientific Writings*, London, 1864.

Linacre, *Essays on the Life and Work of Thomas Linacre c.1460–1524*, Francis Maddison, Margaret Pelling, and Charles Webster (Eds.), Oxford, 1977.

Lloyd, G.E.R., 'Aspects of the relationship between Aristotle's psychology and his zoology', in *Essays on Aristotle's* De Anima, Nussbaum and Rorty (Eds.), Oxford, 1992, 147–167.

Lloyd, Geoffrey, *Aristotle. The Growth and Structure of his Thought*, first publ. 1968 ed. Cambridge, 1980.

Lower, Richard, *A facsimile edition of Tractatus de Corde item De motu & colore sanguinis et Chyli in eum transitu, by Richard Lower, M.D., London 1669*, Prefaced by an Introduction and Tr. K.J. Franklin, (volume 9 of *Early Science in Oxford*, R.T. Gunther (Ed.)), Oxford, 1932.

Lower, Richard, *Vindicatio*, see Dewhurst.

Meyer, Arthur William, *An Analysis of the De Generatione Animalium of William Harvey*, Stanford, 1936.

Minelli, Alessandro, *The Botanical Garden of Padua 1545–1995*, Venice, 1995.

Nussbaum, Martha C. and Rorty, Amélie Oksenberg, Eds., *Essays on Aristotle's* De Anima, Oxford, 1992.

*Bibliotheca Anatomica, Medica, Chirurgica. &c, Containing a Description of the Several Parts of the Body*, Nutt, John and Morphew, John (Eds.), London, 1711–1714; a copy of this rare part-work is in the Wellcome Library in London.

O'Malley, C.D., Poynter, F.N.L., and Russell, K.F., *William Harvey: Lectures on the Whole of Anatomy*. An annotated translation of *Praelectiones Anatomiae Universalis*, Berkeley, University of California Press, 1961.

Ogle, William, translator, *Aristotle on the Parts of Animals*, London, 1882.

Pagel, Walter, *William Harvey's Biological Ideas: Selected Extracts and Historical Background*, New York, 1967.

Pagel, Walter, *New Light on William Harvey*, Basel, 1976.

Pellegrin, Pierre, *Aristotle's Classification of Animals. Biology and the Conceptual Unity of the Aristotelian Corpus*, tr. Anthony Preus, Berkeley, 1986; French original, Paris, 1982.

Pliny Secundus, *The Historie of the World: Commonly called the Natural Historie of C. Plinius Secundus*, Translated into English by Philemon Holland Doctor of Physic, London, 1664 (first published 1601).

*Pliny the Elder. Natural History. A Selection*, Healy, John F., (Ed.), London, 1991.

Pra, Mario Del, 'Una "Oratio" Programmatica di G. Zabarella', *Rivista Critica di Storia della Filosofia*, 21, 1966, 286–290.

Randall, John Hermann Jr., *The Career of Philosophy from the Middle Ages to the Enlightenment*, Columbia, 1962.

Rippa Bonati, Maurizio and Pardo-Tomás, José, Eds., *Il Teatro dei Corpi:* Le Pitture Colorate d'Anatomia *di Girolamo Fabrici d'Acquapendente*, Milan, 2004.

Robinson, H. Wheeler, 'Blood', in *The Encyclopaedia of Religion and Ethics*, ed. Hastings, J., 1909.

Rossetti, Lucia, *The University of Padua: An Outline of its History*, Triest, 1983 (first ed. Milan, 1972).

Schmitt, Stéphane, *Aux Origines de la Biologie Moderne: L'anatomie Comparée d'Aristote à la Théorie de l'évolution*, Paris, 2006.

Semenzato, Camillo, Ed., *The Anatomy Theatre [at Padua.]: History and Restoration*, Padua, 1995.

Shaw, James Rochester, 'Models for cardiac structure and function in Aristotle', *Journal of the History of Biology*, 5, 1972, 355–388.

Stolberg, Michael, 'Learning anatomy in late sixteenth-century Padua', *History of Science*, 56, 2018, 381–402.

Thompson, D'Arcy Wentworth, *On Aristotle as a Biologist: With a Proœmion on Herbert Spencer. Being the Herbert Spencer Lecture Delivered before the University of Oxford, on February 14, 1913,* Oxford, 1913.

Tosoni, Pietro, *Della Anatomia degli Antichi e della Scuola Anatomica Padovana Memoria*, Padua, 1844 (facs., 1995).

Twysden, John, *Medicina Veterum Vindicata*, London, 1666.

Wear, Andrew, 'William Harvey and the "way of the anatomists"', *History of Science*, 21, 1983, 223–249.

Webster, Charles, 'William Harvey's conception of the heart as a pump', *Bulletin of the History of Medicine*, 39, 1965, 508–517.

Webster, Charles, 'Harvey's *De generatione*: its origins and relevance to the theory of circulation', *British Journal for the History of Science*, 3, 1967a, 262–274.

Webster, Charles, 'The College of Physicians: "Solomon's House" in Commonwealth England', *Bulletin of the History of Medicine*, 41, 1967b, 393–412.

Wharton, Thomas, *Thomas Wharton's Adenographia, first published at London in 1656. Translated from the Latin by Stephen Freer with an Historical Introduction by Andrew Cunningham*, Oxford, 1996.

White, J. S., 'The 1653 English edition of *De Motu Cordis*, shown to be Harvey's vernacular original and revealing crucial aspects of his pre-circulation theory and its connection to the discovery of the circulation of the blood'. *History and Philosophy of the Life Sciences*, 21, 1999, 65–91.

Whitteridge, Gweneth, Ed., *The Anatomical Lectures of William Harvey. Praelectiones Anatomiae Universalis* [and] *De musculis*, London, 1964.

Whitteridge, Gweneth, *William Harvey and the Circulation of the Blood*, London, 1971.

Wilkin, Rebecca M., 'Figuring the dead Descartes: Claude Clerselier's *Homme* de René Descartes', *Representations*, 83, 2003, 38–66.

Willis, Robert, *William Harvey: A History of the Discovery of the Circulation of the Blood*, London, 1878.

Willis, Thomas, *De febribus* (1659), published in his *Diatribae duae Medico-philosophicae, quorum prior agit De Fermentatione sive De motu intestino particularum in quovis corpore. Altera De Febribus, sive De motu earundem in sanguine Animalium*, London, 1659. These two works also appear in *The remaining medical works of that famous and renowned physician Dr Thomas Willis*, tr. Samuel Pordage, and published in London in 1681.

Woodbridge, Frederick J. E., *Aristotle's Vision of Nature*, Randall, J. H. R. Jr., (Ed.), Columbia, 1965; the text dates from 1930.

# Index

For Product Safety Concerns and Information please contact our EU
representative  GPSR@taylorandfrancis.com
Taylor & Francis Verlag GmbH, Kaufingerstraße 24, 80331 München, Germany

www.ingramcontent.com/pod-product-compliance
Lightning Source LLC
Chambersburg PA
CBHW060305220326
41598CB00027B/4245